Entrepreneurship and Work in the Gig Economy

The crisis caused by COVID-19 poses a major challenge for the global economy and business. It has been a test for economic resilience, and how this crisis will affect business activities will be determined by their competitiveness. Only firms that have succeeded in reorienting and quickly adapting to the emerging crisis have continued without interruption in their operations, thus demonstrating their flexibility and high level of resilience.

Research shows that companies driving the gig economy celebrate the benefits: flexibility, additional income, freedom and other various opportunities for workers. They require entrepreneurial digital skills that improve their competitiveness and benefit not only themselves but the economy as well. Therefore, digital competencies are becoming a significant resource and precondition for employment, and it is essential to promote digital entrepreneurial skills, introducing them into education programs through different forms of education. This book stresses and explores the importance of synergy between industry and virtual universities. Considering the importance of these issues, this book draws together literature on globalization and small and medium enterprise development and internationalization from disparate sources into a cohesive body of work, which traces the evolution of entrepreneurship and our understanding of the topic. It investigates the impact of digitalization on creating job opportunities in the labor market.

Covering social, economic and psychological approaches to the most current topics, this book will be a useful framework to new theories and practices that are emerging to challenge conventional wisdom. It will be of interest to scholars, upper-level students and researchers involved in the field of entrepreneurship.

Mirjana Radović-Marković is professor and principal research fellow at the Institute of Economic Sciences, Belgrade, Serbia.

Borislav Đukanović is professor at the University Donja Gorica, Podgorica, Montenegro.

Dušan Marković is professor of applied studies at the Belgrade Academy of Business and Art, Belgrade, Serbia.

Arsen Dragojević is a researcher at Ipsos Strategic Marketing in Belgrade, Serbia.

Routledge Focus on Business and Management

The fields of business and management have grown exponentially as areas of research and education. This growth presents challenges for readers trying to keep up with the latest important insights. Routledge Focus on Business and Management presents small books on big topics and how they intersect with the world of business research.

Individually, each title in the series provides coverage of a key academic topic, whilst collectively, the series forms a comprehensive collection across the business disciplines.

Digitalised Talent Management
Navigating the Human-Technology Interface
Edited by Sharna Wiblen

Human Resource Management in the Indian Tea Industry
Nirmal Chandra Roy and Debasish Biswas

Entrepreneurship for Rural Start-ups
Lessons and Guidance for New Venture Creation
Edited by Gloria Jiménez-Marín, Alejandro López Rodríguez, Miguel Torres García and José Guadix Martín

Entrepreneurship and Work in the Gig Economy
The Case of the Western Balkans
Mirjana Radović-Marković, Borislav Đukanović, Dušan Marković and Arsen Dragojević

Innovation Research in Technology and Engineering Management
A Philosophical Approach
Marc J. de Vries

For more information about this series, please visit: www.routledge.com/Routledge-Focus-on-Business-and-Management/book-series/FBM

Entrepreneurship and Work in the Gig Economy
The Case of the Western Balkans

Mirjana Radović-Marković, Borislav Đukanović, Dušan Marković and Arsen Dragojević

NEW YORK AND LONDON

First published 2021
by Routledge
52 Vanderbilt Avenue, New York, NY 10017

and by Routledge
2 Park Square, Milton Park, Abingdon, Oxon, OX14 4RN

Routledge is an imprint of the Taylor & Francis Group, an informa business

© 2021 Mirjana Radović-Marković, Borislav Đukanović, Dušan Marković and Arsen Dragojević

The right of Mirjana Radović-Marković, Borislav Đukanović, Dušan Marković and Arsen Dragojević to be identified as authors of this work has been asserted by them in accordance with sections 77 and 78 of the Copyright, Designs and Patents Act 1988.

All rights reserved. No part of this book may be reprinted or reproduced or utilised in any form or by any electronic, mechanical, or other means, now known or hereafter invented, including photocopying and recording, or in any information storage or retrieval system, without permission in writing from the publishers.

Trademark notice: Product or corporate names may be trademarks or registered trademarks, and are used only for identification and explanation without intent to infringe.

Library of Congress Cataloging-in-Publication Data
A catalog record for this book has been requested

ISBN: 978-0-367-72577-8 (hbk)
ISBN: 978-0-367-72579-2 (pbk)
ISBN: 978-1-003-15540-9 (ebk)

Typeset in Times New Roman
by Apex CoVantage, LLC

Contents

List of Tables x
List of Figures xi
Preface xiii
Acknowledgment xv

PART I
Theoretical Perspectives of Globalization 1

1 **Global Business Flows** 3
   MIRJANA RADOVIĆ-MARKOVIĆ

   1.1 Introduction 3
   1.2 Basic Concept and Different Dimensions of
       Globalization 4
       1.2.1 Globophilia and Globophobia 4
   1.3 Conclusion 7
   References 8

PART II
Globalization and Entrepreneurship 11

2 **Entrepreneurship: Past, Present and Future** 13
   MIRJANA RADOVIĆ-MARKOVIĆ

   2.1 The Development of Entrepreneurship Through
       History and Shift of Globalization to a More
       Digital Form 13
       2.1.1 Motivation Factors When Starting a
             Business 17

2.1.2 Global Entrepreneurship Trends 19
    2.1.2.1 Global Entrepreneurship Index 20
    2.1.2.2 Business Environment 22
2.1.3 Conclusion 23
References 25

PART III
Digital Entrepreneurship 29

3 Enterprise Digital Transformation Toward Network Virtualization 31
MIRJANA RADOVIĆ-MARKOVIĆ

3.1 The Digital Enterprise Concept and Its Virtualization 31
    3.1.1 Introduction 31
    3.1.2 Company Networking 32
    3.1.3 Employee Agility 33
        3.1.3.1 Features of Employee Agility 34
    3.1.4 Different Roles of Virtual Teams in the Realization of the Virtual Companies' Functions 34
    3.1.5 Importance of Communication in Virtual Enterprises 36
        3.1.5.1 Areas of Electronic Communication 37
    3.1.6 The Complexity of Putting Together Project Teams 38
3.2 Conclusion 38
References 38

4 Exploring the Synergistic Potential in Virtual University and Virtual Enterprise 40
DUŠAN MARKOVIĆ

4.1 Introduction 40
    4.1.1 The Concept of Distance Learning 40
    4.1.2 Prospects for the Development of Education via the Internet 41

4.2 Creative Education as a Strategic Form of
    Education 44
    4.2.1 Encouraging Individuality in Education 45
    4.2.2 Development of an Education Strategy Based
          on the Choice of Teaching and Learning 46
    4.2.3 The Effects of Educational Strategies to
          Encourage Individuality 47
    4.2.4 The Impact of Information Technologies on
          Education 48
4.3 Integration of a Virtual University and a Virtual
    Enterprise 50
    4.3.1 Modeling the Integration of a Virtual University
          and a Virtual Enterprise 51
4.4 Conclusion 57
References 58

PART IV
Changing Nature of Work in the Digital Era   61

5 **The Transformation of Work in a Global Knowledge
  Economy**   63
  MIRJANA RADOVIĆ-MARKOVIĆ

  5.1 The Impact of Globalization on New Forms of Labor
      and Employment 63
      5.1.1 Types of Platform Companies 64
  5.2 Changed Manager Role 65
  5.3 Strengthening Digital Entrepreneurial Skills in
      the Western Balkans: Long-Term and Short-Term
      Goals 65
  5.4 Global Gig Economic Index 66
      5.4.1 Representation of the Gig Economy in the
            Western Balkans: Case Study of Montenegro 67
            5.4.1.1 Research Method 68
            5.4.1.2 Key Findings and Discussion 68
  5.5 Conclusion 72
  References 73

PART V
Organizational and Entrepreneurial Resilience 75

6   Resilience and Entrepreneurship   77
MIRJANA RADOVIĆ-MARKOVIĆ

    6.1 Organizational Resilience 77
        6.1.1 The Modern Concept of Resilience 78
    6.2 Resilience and Competitiveness 79
    6.3 Regional Resilience in Times of a Pandemic Crisis:
        The Case of Montenegro 81
        6.3.1 Research Method 82
            6.3.1.1 Key Findings and Discussion 82
    6.4 Conclusion 88
    References 88

7   The Resilient Entrepreneur   90
ARSEN DRAGOJEVIĆ

    7.1 Entrepreneurial Resilience 90
        7.1.1 Methodology 91
        7.1.2 Key Findings and Discussion 92
    7.2 Conclusion 94
    References 95

PART VI
Economic and Social Impact of Global Ecosystem on Progress in Montenegro 97

8   Economic Prosperity in the Western Balkans and Montenegro   99
MIRJANA RADOVIĆ-MARKOVIĆ

    8.1 The Economic Prospects and Further Prosperity 99
    8.2 Economic Prosperity of Montenegro 99
        8.2.1 The Legatum Global Prosperity Index 100
            8.2.1.1 The Pillars of Montenegro's Economic Progress 101
    8.3 Conclusion 102
    References 103

9 Measuring Quality of Life — 105

BORISLAV ĐUKANOVIĆ AND ARSEN DRAGOJEVIĆ

9.1 The Social Prospects: Quality of Life Indicators in Montenegro 105
   9.1.1 Introduction 105
   9.1.2 Theoretical Concept of Research 105
   9.1.3 Research Method 109
   9.1.4 Key Findings 110
   9.1.5 Conclusion 131
9.2 Social Prospects of Montenegro: Indexes 133
   9.2.1 Democracy Index 134
   9.2.2 Corruption Perception Index 137
   9.2.3 Global Health Security Index 138
   9.2.4 Social Progress Index 140
   9.2.5 Global Multidimensional Poverty Index 142
   9.2.6 Global Gender Gap Index 144
   9.2.7 Human Freedom index 145
9.3 Conclusion 147
References 149

10 Measuring Professional Life — 152

BORISLAV ĐUKANOVIĆ

10.1 Professional Life in Montenegro: Research 152
   10.1.1 Research Method 152
   10.1.2 Key Findings 153
10.2 Conclusion 156
References 157

About the Authors 158
Index 159

Tables

1.1	Overview of globalization theories: Classical and contemporary theories	5
2.1	Global entrepreneurship index (GEI)	21
2.2	Business environment indicators	23
2.3	Ranking of countries by four basic indicators, 2019	23
4.1	The input, output and control parameters of the integration of the virtual university and the virtual enterprise	52
6.1	Global resilience index for the Western Balkans, 2019	80
6.2	Demographic structure by age	83
6.3	Level of education	83
6.4	Do you expect that the COVID-19 crisis will reflect on SMEs?	84
6.5	The worst scenario if the company has no business continuity plan	85
6.6	How many times did your company stop working until the crisis?	86
6.7	The most effective recovery strategy	86
9.1	Correlations between strata affiliation and satisfaction of respondents in 10 areas of family life	114
9.2	Matrix of factor saturations of value preference items	122
9.3	Democracy index for Western Balkan countries and country type	135
9.4	Factors of democracy index for selected countries	135
9.5	Corruption perception index for Western Balkan countries	138
9.6	Social progress index and its two main factors for Western Balkan countries, 2019	141
9.7	Global gender gap index and its two main factors for Western Balkan countries, 2020	145
9.8	Freedom index for Western Balkan countries, 2019	146
10.1	Problems at work	155
10.2	Life plans of respondents in the next five years	156

Figures

2.1	Western Balkan countries' ranking by Doing Business List, 2019	21
2.2	Impact of globalization on entrepreneurship	22
3.1	Virtual companies	32
3.2	Elements of employees' agility	33
4.1	Percentage of the population in the European Union using the internet	41
4.2	Percentage of the population in Europe using the internet to search for training, education information or course offers	42
4.3	The first contextual level of VU and VE integration	51
4.4	Decomposition of the synergy of the virtual university and the virtual enterprise	53
4.5	Decomposition of the integrated synergy system (VU+VE)	54
5.1	Countries ranked by highest revenue from freelancers in the world	67
5.2	Structure of employees in the gig economy by level of education	69
5.3	The kind of jobs in the gig economy	69
5.4	In what aspects of life, the gig economy has positively affected	70
5.5	In what aspects of life, the gig economy has negatively affected	71
5.6	How can the position of freelancers be best improved?	71
5.7	How working in a gig economy can reduce brain drain	72
6.1	Global resilience index	79
6.2	The impact of economic resilience on competitiveness	81
6.3	Do you expect that the COVID-19 crisis will reflect on SMEs?	84
6.4	Worst scenario	85

6.5	How can the effectiveness of a recovery plan be determined?	87
6.6	Firms who have stopped working in a case of impossibility to respond to disaster (in percent)	87
8.1	Legatum prosperity index, 2018–2019	102
8.2	The progress of the countries of the region according to Legatum's pillar prosperity index, 2019	103
9.1	Democracy index of Montenegro for period 2006–2019	136
9.2	Corruption perception index of Montenegro for period 2012–2019	138
9.3	Social progress index and its main factors for Montenegro for 2014–2019	142
9.4	Main four factors of global gender gap index for Montenegro, 2020	143
9.5	Global gender gap index of Montenegro for the period 2015–2020	145
9.6	Freedom index of Montenegro for 2015–2020	147

Preface

The current situation with the crisis caused by COVID-19 poses a major challenge for the global community. Many scientists are predicting the end of globalization, whereas others are announcing a whole new way of managing global business. One can also see comparisons of the new pandemic crisis with that of the second half of the 19th century, when yellow fever and other viruses raged, changing the flow of investment and trade. But even during this period and later during the Great Economic Depression, global cash flows continued, as did the expansion of networking in science, health and other areas of international connectivity. According to some historians (Rosenberg, 2014), these global flows have even accelerated during major crises. Accordingly, there is a dilemma whether this experience can be applied in the present circumstances? The issue of the position of small countries in such changed conditions of economic and social life, and especially of the Western Balkan countries belonging to one of the least developed regions of Europe, is particularly raised as well. First of all, the aforementioned is referring to their economic and social survival, because as small economies, they are highly dependent on international trade and foreign investment. In addition, a large disruption in the normal mode of operation and closure of a large number of jobs caused by the crisis raised the issue of new modalities. Current developments have shown that only those firms that have succeeded in reorienting and quickly adapting to the emerging crisis have continued without interruption in their operations, thus demonstrating their flexibility and high level of resilience. Activating entrepreneurial digital skills improves the competitiveness of businesses that benefit not only organizations but the economy as a whole. Companies that have a developed digital infrastructure and that use digital technologies are usually the most competitive. In contrast, the lack of digital investment and infrastructure can put companies at a disadvantage. Therefore, digital competencies are becoming a significant resource and precondition for employment both globally and in the Western Balkan countries. In the absence thereof, in the near future there could be an increase in structural mismatch in the labor market. Those with the lowest levels of digital skills would be most affected as well as those who are least willing to update their skills. Accordingly, it is necessary to promote digital entrepreneurial skills and introduce them into education programs through different forms and levels of education.

Considering the importance of these issues, they will be included in the framework in our study on the example of Montenegro as the smallest country in the region, comparing it with other countries in the immediate area. This would not only point out the similarities and differences that exist among them but also analyze the most current topics that have not hitherto been considered in scientific circles in such an integral and coherent way. This would at the same time give guidance to economic and social policy-makers not only to overcome the crisis more quickly but also for further development.

Mirjana Radović-Marković, PhD
June 2020

Acknowledgment

'The Montenegrin Academy of Sciences and Arts is specifically dedicated to examining the economic, cultural and political impacts of globalization on small countries. To that end, it has organized several international scientific conferences and helped conduct a series of scientific research.'

This research was conducted with the help of the Montenegrin Academy of Sciences and Arts, for which we owe them great gratitude.

Mirjana Radović-Marković, Borislav Đukanović,
Dušan Marković and Arsen Dragojević

Part I
Theoretical Perspectives of Globalization

1 Global Business Flows

Mirjana Radović-Marković

1.1 Introduction

Globalization is shaping our economic, social and political realities in numerous ways. Above all, it has a major impact on the development of global finance and financial markets, the dissemination of knowledge through improved communication conditions, the expansion of multinational firms and the decentralization of economic activities within and between firms, the development of new forms of employment and job opportunities and the reduction of barriers to trade and investment. When organizations integrate with one another, they cross local boundaries and simultaneously contribute to the economies of several countries. Based on the opinion of many scholars, globalization has interrupted previously strong local and regional economic identities of countries (Sánchez, 2010; Radović-Marković, 2019; Alishahi, Refiei, and Souchelmaei, 2019). Increased integration has led to an increased need for a new type of social, political and legal regime. With this in mind, it is logical that the process of globalization becomes an inevitable starting point in all economic analyses at the national level, because modern economic development becomes impossible without accepting the trends in the global market. In accordance with these requirements, specific objectives of this part of the study have been formulated, relating to:

1. Determining the degree of integration of Montenegro and other Western Balkan countries into the global business flows and their positioning in the global market.
2. Exploring the impact of globalization on economic development; competitiveness; innovation; entrepreneurship development and new forms of education, work and employment as well as the resilience of states and enterprises to external and internal shocks (contagions, earthquakes, climate changes, etc.).

1.2 Basic Concept and Different Dimensions of Globalization

The concept of globalization is widely set and as such encompasses a number of disciplines analyzed from different perspectives. For the purposes of this study, the focus will be on how globalization reaches every aspect of economic and social life. This cause-and-effect relationship is often articulated without empirical evidence, which leaves a correlation between globalization and its effect without any measure. However, without measuring the effects of globalization, valid conclusions cannot be drawn, but they remain at the level of arbitrariness of interpretation.

The economic aspect of the KOF globalization index measures the flow of goods, capital and services over long distances as perceptions that accompany market exchanges. In addition to actual trade flows and foreign investment, it also covers the extent to which a country restricts capital and trade flows. On the other hand, the social dimension of globalization measures the spread of ideas and information.

Given the large number of theories related to the phenomenon of globalization, they need to be systematized into theories that come from the views of classical scholars and those of contemporary thinkers. The difference between the two approaches is that classical scholars have discussed cultural and sociological views, whereas contemporary scholars have emphasized the economic and political aspects of the phenomenon of globalization (Table 1.1).

1.2.1 Globophilia and Globophobia

Globalization is one of the most widely debated issues of the present age. Generally speaking, the debate involves those who are supporters of globalization or 'globophiles', as opposed to those who do not support global processes or 'globophobes'. Namely, on the one hand, there are 'globophiles' who believe that globalization connects the world and brings many economic benefits to countries that are well integrated into international business flows. For them, globalization is a continuation of modernization and a driver of progress, increased wealth, freedom, democracy and happiness. Further, it encourages more trade and increased capital inflows among nations. For example, businesses may seek their financing from foreign banks offering the most competitive interest rate (Bhagvati, 2005).

Our recent research has shown that the effect of economic globalization depends on the GDP level of a country, regardless of its size (Radović-Marković and Tomaš, 2019). The relationship between economic globalization and economic growth is important especially for economic policies (Samimi and Jenatabadi, 2014). Specifically, the impact of globalization on

Theories	Authors	Characteristics	Sources:
Modernization theory	Max Weber and Talcott Parsons	It is based on economic development, political and economic stability and democratization of society	Nash, K. (2007). *Contemporary Political Sociology: Globalization, Politics and Power*. Oxford: Wiley-Blackwell
The theory of postmodernism	Jean-François Lyotard, Jacques Derrida and Fredric Jameson	Postmodernism is directly connected with the process of globalization. The representatives of this theory emphasize the importance of knowledge. They also focus their attention on the economic and cultural aspects of globalization.	Nozari, H.A. (2005). *The Formulation of Modernity and Post-modernity*. Tehran, Iran: Jahaneghsh Press. Harvey, D. (1989). *The Condition of Postmodernity: An Enquire into the Condition of Cultural Change*. Oxford: Wiley-Blackwell.
Hyperglobalist theory	Thomas Friedman	The hyperglobalist approach is based on the view that the world has entered a 'global age', legalizing the dominance of 'global capitalism'. The logic of the hyperglobalist stance is based on 'neoliberal capitalism'.	Tikly, L. (2001). Globalisation and education in the postcolonial world: Towards a conceptual framework. *Comparative Education*, 37(2): 151–171. Held, D. (2004). Democratic accountability and political effectiveness from a cosmopolitan perspective. *Government and Opposition*, 39(2): 364–391.
Dependency theory	Raul Prebisch	Globalization is linked to multinational corporations, international commodity markets, foreign aid and foreign investment.	Ferraro, V. (2008). Dependency theory: An introduction. In Giorgio Secondi (Ed.), *The Development Economics Reader*. London: Routledge.
World-systems theory	Immanuel Maurice Wallerstein	This theory places emphasis on capital-intensive production and trade dominance. According to this theory, the world is divided into central countries, semi-peripheral countries and peripheral countries.	Robinson, W. (2011). Globalization and the sociology of Immanuel Wallerstein: A critical appraisal. *International Sociology*, 26(6): 723–745. Available at: http://citeseerx.ist.psu.edu/viewdoc/download?doi=10.1.1.918.9776&rep=rep1&type=pdf (accessed on 21 March 2020).

Source: Author.

the economic growth of countries could be altered by a series of complementary policies such as improving human capital and the financial system. The effect of complementary policies is very important, as it helps countries be successful in the globalization process. Although Nobel laureate Krugman (1993) agrees that improving the financial system can contribute to the success of the globalization process, he emphasized that international financial integration is not a major engine of economic development.

By increasing market size through globalization, countries can benefit from economies of scale, lower costs of exploration and knowledge dissemination. In addition, some researchers argue that the effects of the impact of globalization on economic growth depend on the economic structure of countries during the globalization process.

The greatest 'globophobes' believe that globalization is a means for developed nations to colonize developing countries through economic control and that these countries are helpless victims of globalization. In support of this claim, they argue that globalization enables developed countries to exploit the natural resources of developing countries and deplete nonrenewable resources. Many have identified globalization as a major contributor to inequality among countries, and seen as particularly responsible are multinational corporations, which are in constant search for cheap labor and high profits (Ritzer, 2007). Specifically, global companies build mutual benefit relationships that cross national boundaries (Armstrong, 2020).

Some economists have singled out an increase in poverty, increase in the fiscal deficit and, above all, more pronounced regional differences in the world as a consequence of global processes (Kalyan, 2000). The worsening of economic inequality has called into question the ability of economic globalization to develop lagging regions (Feffer, 2020). Particularly, according to a number of scholars, globalization has a detrimental effect on economic growth in countries with weak institutions and political instability (Borensztein, De Gregorio, and Lee, 1998).

Complementing the negative view of global processes, 'globophobes' claim that globalization produces the undermining of democracy, cultural homogenization and increased destruction of natural species and the environment (Appadurai, 1990).

Bello (2007), one of the leading critics of globalization, suggested that countries place emphasis on trade at national rather than global levels to protect local economies, improve quality of life and support regional rather than global institutions.

The economic crisis of 2008 showed a close link between the crisis and the decline in the level of economic globalization in the world. Accordingly, the question arises whether the new economic crisis of 2020 caused by the coronavirus will again slow down the processes of economic globalization or even lead to the end of globalization, as some predictions have emerged. Such statements are corroborated by the fact that when the world is faced with a pandemic and the threat of financial collapse in 2020, each country for itself has been forced to seek a solution to closing their national borders. Further, as it moves around the world, the coronavirus threatens the circular system of globalization, dramatically reducing the international flow of money, goods and people. This puts global economic relations at risk. In addition, as a result of the impact of the coronavirus, there is an increasing demand from Western Europe to reduce dependence on third-world suppliers.

In line with these requirements, and given that global economic integration has slowed over the last decade, culminating in the emergence of the COVID-19 crisis, experts are increasingly suggesting that globalization has passed its peak and that there has to come about a gradual deglobalization (Feffer, 2020; Herrero, 2020).

1.3 Conclusion

By the end of March 2020, the pandemic had wiped out $3 billion from global stock exchanges and threatened the future of millions of small businesses around the world, along with the opportunity for a huge number of employees to work and make money (Karabell, 2020). Accordingly, one group of experts believes that globalization is losing its importance in these circumstances, which will in turn lead to deglobalization. Despite this statement, we believe that even in a crisis like this, globalization can play an important role and become a source of global economic stability, first of all, due to the fact that national economies are connected by 'connective tissue' with global companies. Consequently, the new economic crisis will make it difficult to halt the long-standing international processes of interconnecting countries in a global context. In other words, it is not possible that new contacts between people and countries can be completely prevented. They will only get their alternatives through digital communication. Thus, digital globalization will have a primacy, especially as it is not vulnerable to a pandemic, and teleworking will receive a huge boost from coronaviruses. It will continue to some extent even when the crisis passes.

Although it is difficult when the pandemic continues to give accurate predictions of what the consequences of this crisis will be, what is certain is the fact that the international community will have to strike a new balance between national and global development priorities. Accordingly, globalization will take on new forms, and it can be expected that the anti-globalist rhetoric will be reduced and mitigated. Finally, it can be concluded that the current crisis is global, but that no globalization crisis has occurred (Armstrong, 2020).

References

Alishahi, A., Refiei, M., and Souchelmaei, H.S. (2019). The prospect of identity crisis in the age of globalization. *Global Media Journal*, 17(32): 1–4.

Appadurai, A. (1990). Disjuncture and difference in the global cultural economy. *Theory, Culture & Society*, 7: 295–310.

Armstrong, R. (2020). *Coronavirus is a Global Crisis, Not a Crisis of Globalisation*. Available at: www.ft.com/content/5e933fce-62bb-11ea-b3f3-fe4680ea68b5/ (accessed on 19 March 2020).

Bello, W. (2007). Globalization in retreat. *Frontline*, 24(1): 57–59.

Bhagvati, J. (2005). In defense of globalization. *International Journal*, 60(2): 592–595.

Borensztein, E., De Gregorio, J., and Lee, J.W. (1998). How does foreign direct investment affect economic growth? *Journal of International Economics*, 45: 115–135.

Feffer, J. (2020). *Will the Coronavirus Kill Globalization?* Available at: www.com mondreams.org/views/2020/03/07/will-coronavirus-kill-globalization (accessed on 21 March 2020).

Ferraro, V. (2008). Dependency theory: An introduction. In Giorgio Secondi (Ed.), *The Development Economics Reader*. London: Routledge.

Harvey, D. (1989). *The Condition of Postmodernity: An Enquire into the Condition of Cultural Change*. London: Routledge.

Held, D. (2004). Democratic accountability and political effectiveness from a cosmopolitan perspective. *Government and Opposition*, 39(2): 364–391.

Herrero, A.G. (2020). *The Pandemic Will Structurally Change the Global Economy More Than We Think*. Available at: https://www.brinknews.com/the-coronavirus-pandemic-will-change-the-global-economy-more-structurally-than-we-think/ (accessed on 25 December 2020).

Kalyan, R. (2000). Conference of economists: A report. *EPW*, XXXV(19): 1612–1613.

Karabell, Z. (2020). *Will the Coronavirus Bring the End of Globalization? Don't Count on It*. Available at: www.wsj.com/articles/will-the-coronavirus-bring-the-end-of-globalization-dont-count-on-it-11584716305/ (accessed on 18 March 2020).

Krugman, P. (1993). International finance and economic development. In A. Giovannini (Ed.), *Finance and Development: Issues and Experience* (pp. 11–24). Cambridge: Cambridge University Press.

Nash, K. (2007). *Contemporary Political Sociology: Globalization, Politics and Power*. Oxford: Wiley-Blackwell.

Nozari, H.A. (2005). *The Formulation of Modernity and Post-modernity*. Tehran, Iran: Jahanneghsh Press.

Radović-Marković, M. (2019). *Jačanje naučne saradnje u oblasti društvenih i ekonomskih nauka izmedju Sjedinjenih Američkih Država i Srbije*. Beograd: Filozofski fakultet (grupa za istoriju) i Institut ekonomskih nauka.

Radović-Marković, M. and Tomaš, R. (2019). *Globalization and Entrepreneurship in Small Countries*. New York: Routledge. ISBN 9780367250751.

Ritzer, G. (2007). *Globalization*. Oxford: Wiley-Blackwell Publishing.

Robinson, W. (2011). Globalization and the sociology of Immanuel Wallerstein: A critical appraisal. *International Sociology*, 26(6): 723–745. Available at: http://citeseerx.ist.psu.edu/viewdoc/download?doi=10.1.1.918.9776&rep=rep1&type=pdf (accessed on 21 March 2020).

Samimi, P. and Jenatabadi, H.S. (2014). Globalization and economic growth: Empirical evidence on the role of complementarities. *PLoS One*, 9(4): e87824. https://doi.org/10.1371/journal.pone.0087824

Sánchez, M.E. (2010). Globalisation and loss of identity. *International Forum of Psychoanalysis*, 19(2): 71–77. https://doi.org/10.1080/08037060903435158(28). Available at: https://www.researchgate.net/publication/233193630_Globalisation_and_loss_of_identity (accessed on 26 December 2020).

Tikly, L. (2001). Globalisation and education in the postcolonial world: Towards a conceptual framework. *Comparative Education*, 37(2): 151–171.

Part II
Globalization and Entrepreneurship

2 Entrepreneurship
Past, Present and Future

Mirjana Radović-Marković

2.1 The Development of Entrepreneurship Through History and Shift of Globalization to a More Digital Form

From a historical perspective, entrepreneurship is one of the oldest economic activities. Namely, discovering and identifying business opportunities for starting a new business have always been important for human life and prosperity as well as for the social and economic community as a whole. One of the key shifts in the history of entrepreneurship (and human history) was the invention of money. Before the invention of money, all entrepreneurship and commerce took place through the barter system. This system of trade had a great influence on the limitations of commerce and entrepreneurship in early history.

Entrepreneurial activities in society are mentioned even about the ancient Greeks. They are found in the works of the Greek philosopher Xenophanes (around 430–354 BC). This early type of entrepreneurship has continued for millennia. Thus, in the Middle Ages, the typical entrepreneur was responsible for the construction of castles, state buildings and churches. From the 13th century, they not only engaged in the construction of churches, castles and similar structures but continued to expand their entrepreneurial activities in construction. However, it was not known at that time what the tasks of the entrepreneurs were and what their significance was for the benefit of the economic and social community. In the mid-18th century, the individual influence of entrepreneurs on changes in the economy and society began to be considered.

It was not until the 20th century that the first comprehensive definition of entrepreneurship was given, with the advent of Schumpeter's (1934) papers, which emphasized the importance of entrepreneurship for fostering innovation and creativity. In this way, entrepreneurship and the theory of economic growth were brought into relation and connected. Schumpeter has repeatedly emphasized that his empirical study of entrepreneurship was a historical endeavor, given that this phenomenon can best be understood through the retrospective of critical elements in the process of industrial and economic changes. Therefore, scientific entrepreneurship research needs to focus not only on entrepreneurs and their firms but also on changes that have occurred over time in the industry, market, society, economy and political system in which they operated, and through an eclectic approach that history can only provide.

Interest among scholars from different scientific disciplines in this area of research in the mid-20th century has contributed to a large number of papers analyzing this issue from an economic, historical and social point of view, but entrepreneurship has also become a part of behavioral and cognitive science as well as management science. Therefore, we cannot consider entrepreneurship research as one coherent field, but rather, as a set of many different subtopics. Consequently, entrepreneurship is an increasingly dynamic area of research.

Entrepreneurship can be considered on an individual, local, national and global level. The analysis of entrepreneurial preferences and ideas, their identification and exploitation, has not lost popularity and continues its continuity even with contemporary scholars (Valdez and Richardson, 2013; Munro, 2003). From a methodological point of view, such studies show some drawbacks, because conclusions are drawn on the basis of the generalization of individual examples.

Already in the late 1960s, scientists showed through their research that there is a close link between entrepreneurship and economic development and that entrepreneurship can be developed through planned efforts. Therefore, the development of entrepreneurship and entrepreneurial activity have become part of the economic development strategies of many countries in the world. More recently, the current trends of the modern business world are closely linked and take place under the influence of the globalization of the world economy. These trends include an increasing number of international corporations, reinforcing the forces of global economic competition, identifying new types of businesses and changing business culture and leadership styles and influencing increasing diversity within the workforce. Globalization has also affected small- and medium-sized enterprises in changing their role, above all, in national economies. Therefore, the impact of globalization on the development of entrepreneurship is the subject of many studies (Knight, 2000; Milner and Kubota, 2005). According to Knight (2000), the more a firm is integrated into global business flows, the better its performance. Also, there is an opinion that in the global economy, comparative advantage is gained by firms that create customer loyalty, regardless of geographical location. Specifically, regardless of size or geographical location, a firm can meet global standards and advance and act as a world-class manufacturer and trader, using its concepts, competencies and networking skills (Pologeorgis, 2019).

Today, due to the adoption of globalization and trade liberalization policies in the world, the business environment has become more competitive. In this context, companies require a clear understanding of the foreign market they are entering into. Thus, there are several components in the business environment that need to be analyzed and strategies developed based on the information gathered. These components include the economic, technological, sociocultural and political-legal environment (Francis and Richard, 2017). Specifically, creating a business strategy and organizational model is highly dependent on different cultures, leadership style, motivation and so on. Therefore, the organization needs clearly defined goals and the ability to have a vision of doing business in the global market (Radović-Marković, Nikitović, Vujičić, and Kasumovic, 2019).

In academic literature, we can distinguish different approaches to the impact of entrepreneurship on economic development. The relationship between entrepreneurship and a country's economic growth over the years has increasingly gained the interest of economists and policymakers. However, whereas some see it as an immediate relationship, others see an indirect type of relationship.

Recognizing the capacities and contributions of women entrepreneurs in our global community is no longer a matter of debate, given that women's entrepreneurship is nowadays one of the main contributing factors to the economic development of many countries. Most women entrepreneurs find it increasingly realistic to start their own businesses and often try to develop them in their family environment.

Ayyagari, Kunt, and Maksimovic (2006) analyzed research done in 80 countries at the enterprise level, which aimed to identify those elements in the business environment that most limited the growth of the enterprise. They found that business growth was directly influenced by a lack of funding, street crime and political instability, whereas taxes and laws had no major impact.

The World Bank study states that "lawmakers should focus more on creating a business environment that allows businesses to enter and exit easily and guarantees entrepreneurs and financiers that their rights and contracts will be respected, rather than directly financing SMEs and supporting a large number of small businesses companies" (Beck and Demirgüc-Kunt, 2004).

In her research, Radović-Marković (2008) tried to define the profile of a successful organization that will be best adapted to the multicultural business environment, its changes and challenges. In doing so, in her studies, she relies on new business imperatives, which call for changes in organizational behavior. Globalization has greatly contributed to this, breaking down previous barriers in the international market. Consequently, constant changes have become a fact of organizational life. The author notes that the operational strategies and organizational structure of many companies do not meet the new requirements of the business, which is reflected in the large number of reasons: a) lack of ability to manage projects, b) insufficient support from managers who insist on change, c) finding unilateral solutions and d) unwillingness to change team members in line with project goals and the like. Based on the research, it is concluded that "market champions" will be organizations that are based on continuous learning—that is, learning organizations gain a competitive advantage from continuous improvement both individually and collectively. In addition, to respond to changing conditions and requirements, organizations need to internationalize and cross local boundaries, regardless of their geographical, organizational and other characteristics.

Calantone and Di Benedetto (2000) give the concept of digital enterprise, which refers to intercompany agility. This concept indicates that the use of modern information and communication technologies integrates resources globally. The spatial arrangement of digital enterprise components is an important and necessary condition for system flexibility. The organization initiating the cooperation determines the most suitable business processes, which are complementary to the business skills of different firms. Accordingly, in a new virtual form of organizational structure, they combine their knowledge and experience, share business costs and jointly enter the market. They are mainly characterized by modularity, heterogeneity and spatial and temporal distance. The main objective of such businesses is to enable participating organizations to rapidly develop their work environment. Meeting common goals is achieved through a range of resources, provided by participating organizations.

When designing a team, special attention is paid to its structure. The structure itself should be so integrated that the abilities, skills and personality traits of team members are complementary. Teams with the same or similar expert profiles have not proven effective in practice (Radović-Marković et al., 2014). In addition, practice has shown that in addition to aptitude and work experience, personality criteria should be included in the criteria for selecting team members. Apart from such traits as energy, persistence, perseverance, tactfulness, cooperativeness, loyalty to the company and so on, the success of the team depends on the ability and expertise of the team members (Vučeković, Radović-Marković, and Marković, 2020).

Further, it is important to establish closer links between businesses, governments and the financial sector in order to jointly design innovative management models for SMEs and promote modern business practices. A large portion of SMEs and companies in the region remains 'family run'. This has implications for companies' ability to innovate and their willingness to increase their number of employees (WEF, 2020).

In the European Commission's new Strategy for the Western Balkans: EU (2018), for the first time because of the region's first steps toward the EU, the digital agenda has become central. Launching the digital agenda is one of six flagship initiatives that the EU will undertake in the coming years to support efforts to transform the Western Balkans. Digital transformation is expected to open new businesses and other forms of work. First of all, it is an opportunity to recruit young educated people who can easily and quickly absorb short-term digital skills programs, which gives them different perspectives on online jobs.

2.1.1 Motivation Factors When Starting a Business

Governments in all countries in the region are increasingly implementing policies to start and promote entrepreneurship. Such policies are often implemented without understanding the motivational and other factors that drive an individual to start a particular type of business (Bosma and Levie, 2020).

Research has shown that there are a number of different motivational factors that influence a business startup and entry into entrepreneurial waters. First of all, there is a desire for autonomy in business and freedom to make business decisions. However, apart from it, there are additional motivations such as the desire to enrich oneself. Also, many are embarking on a business, driven by the intent to meet business challenges. There are also many who can only get a job in this way.

It should be emphasized that there are differences between women and men in terms of personal motivation to become entrepreneurs. Many scholars have tried to create a typology of women entrepreneurs. One of the most accepted is the one that classifies women entrepreneurs into the following six groups (Radović-Marković, 2007a):

- young women who do not have a clearly defined business goal but are starting a business to get a job;
- young women who have a clear goal in business and who make long-term business plans. However, they enter the business without enough knowledge and experience and try to supplement this with adequate training programs;
- women who, when setting up their businesses, are guided by their business ambitions and long-term plans as the previous group of women but are much older and more experienced than they are. Most often,

they have a lot of business experience and no family, so they are maximally committed to their business ambitions;
- women who are trying to reconcile their personal responsibilities with their business responsibilities and are therefore looking for an adequate solution that will give them more flexibility and ability to reconcile this dual role;
- women who have lost their jobs and are of very low educational attainment and poor financial status and are therefore forced to look for some new ways to make money;
- women who come from entrepreneurial families who are expected to continue the family tradition.

A review of the academic literature shows that women are more oriented toward achieving their business goals than men, whereas men are more often motivated by good earnings. Recent research has shown that in 34 out of 50 economies, men more than women had as the main motive to continue the family tradition (Bosma and Levie, 2020).

It should be emphasized that women in different economic and cultural backgrounds may have different motives for entering entrepreneurial waters. However, what is common to all of them is the fact that enterprising women and their businesses make a major contribution to the socioeconomic development of their countries through the generation of new jobs and job creation (Radović-Marković, 2015). Therefore, it is necessary to have a better understanding of the importance of women entrepreneurship through the analysis of the characteristics of the businesses they start, the problems they face in their entrepreneurial path, the opening of opportunities and opportunities for their empowerment and the like.

According to Bosma and Levie (2020), by understanding why entrepreneurs start their own businesses, policymakers may be able to reconcile the work being done by different entrepreneurs with national priorities and remove the obstacles entrepreneurs face in achieving their goals.

As there are differences in terms of motivation, so there are also differences in the choice of the type of business among potential entrepreneurs. The choice of type of business depends on several factors, such as (Radović-Marković, 2007b, 25):

- on the entrepreneur's goals in starting a business (business development, benefits in taxation, independence in business, etc.);
- on the dynamics of business development—slowly (part-time engagement) or fast (full-time engagement);
- on the kind of the technological context of the business (high versus low technology);

- on the organizational structure of the company and the form of ownership—partnership, franchising, corporate ownership and so on.

The social system, cultural milieu, political and other conditions under which an entrepreneur starts a new business has an impact on whether or not a new business will be successful. Specifically, the conditions for starting new businesses vary and depend on several factors: from national and geopolitical frameworks to industrial, national and cultural ones. These differences can be taken into account when explaining the difficulties faced by policymakers to accelerate the development of new businesses and entrepreneurial activities.

2.1.2 Global Entrepreneurship Trends

Contemporary entrepreneurship development trends are closely linked to the continuing globalization of the economy. These trends include an increasing number of international corporations, reinforcing the forces of global economic competition, identifying new types of businesses, changing business culture and leadership styles and influencing the increasing diversity within the workforce. A strong sector of small businesses and entrepreneurship is mainly associated with a strong economy (Desai, 2009).

Globalization also presents new challenges for SMEs, leading them to integrate the idea of global change into their business strategies.

Because the impact of many factors resulting from globalization is different for SMEs compared to multinational companies, it is necessary to identify the most important factors and determine their effect.

One group of researchers highlighted several of the most important factors of globalization that directly influence SMEs (Deo, 2013; Gillie, 2018):

1. The rise of cheap communication and transportation.
2. The use of new technologies, which has an important influence on how companies communicate globally and introduce their products and services to the global market.
3. The internationalization of business depending on financial and economic integration.

The conditions that lead to entrepreneurial activities range from personal and cultural to institutional and are influenced by the level of business innovation, the diversity of offers (products and services) and individual entrepreneurial efforts (Wennekers and Thurik, 1999). Globalization rewards companies that are innovative and competitive, regardless of the size of the organization and its country of origin (Radović-Marković and Tomaš, 2019).

The main drivers of globalization in modern conditions—namely, the development of information technology and transport—are canceling out the basic handicaps of small- and medium-sized enterprises, which are related to their inability to communicate effectively with distant markets. Due to the availability of the market in terms of simple and efficient communication and exchange of information with its players, as well as easier and cheaper ways to overcome spatial and other geographical barriers, SMEs are increasingly able to become equal participants in economic activities in the most distant international markets (Radović-Marković, Brnjas, and Simović, 2019).

According to the Global Entrepreneurship Monitor (GEM, 2018/2019 Global Report), four contemporary forms of entrepreneurial activity stand out in contemporary conditions globally:

- 'Solo' entrepreneur.
- Employee entrepreneurial activities.
- Family entrepreneurship.
- 'Gig'—entrepreneurship or entrepreneurship based on internet platforms.

Our research in this study focused on 'gig'—entrepreneurship that uses innovative technology, disrupts existing business models and has a global outlook. We have focused on these types of companies because of their potential for rapid growth and because their needs are understood differently from those of traditional industries.

2.1.2.1 Global Entrepreneurship Index

The global entrepreneurship index (GEI) is an indicator of the quality of entrepreneurship, especially related to the effects of entrepreneurship and innovation, driven by individual and institutional factors. The GEI measures the intensity of the impact of various factors on the development of entrepreneurship in a country. GEI consists of three subindices: 1) entrepreneurial attitudes, 2) entrepreneurial skills and 3) entrepreneurial aspirations (Acs, Szerb, Autio, and Lloyd, 2017). This means that GEI is a three-component index, which takes into account different aspects of the entrepreneurial ecosystem (Radović-Marković and Tomaš, 2019). At the same time, it is an indicator of the quality of entrepreneurship, especially related to the effects of entrepreneurship and innovation, which are conditioned by individual and institutional factors (Table 2.1).

Table 2.1 shows the differences with respect to the GEI for the Western Balkan countries for the period 2015–2019.

Table 2.1 Global entrepreneurship index (GEI)

	2015	2016	2017	2018	2019
Serbia	30.60	30.90	23.13	26.35	28.60
N. Macedonia	37.10	36.60	28.74	29.15	23.0
Montenegro	39.10	37.50	30.21	31.19	31.80
B&H	28.90	28.60	19.94	20.71	19.50
Albania	30.60	30.0	22.97	24.21	22.50

Source: Knoema, 2019.

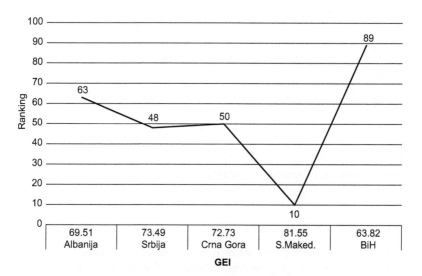

Figure 2.1 Western Balkan countries' ranking by Doing Business List, 2019
Source: Author created, according to World Bank Data (2019a).

Observed for 2019, according to the Doing Business List, Northern Macedonia was the highest-ranked country (10th place) among 137 countries covered. Serbia and Montenegro are closely ranked (48th and 50th position, respectively) and Bosnia and Herzegovina (B&H) is ranked the worst (89th place) (Figure 2.1).

Taking into account the GEI and the globalization index, our analysis showed a moderate correlation between the level of globalization and the level of entrepreneurship for Montenegro and the countries in the region—that is, the correlation is R2 = 0, 4012 (Figure 2.2).

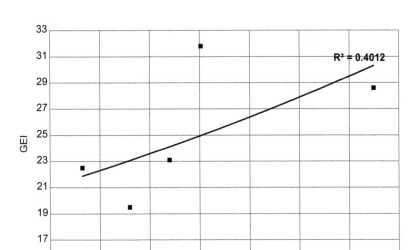

Figure 2.2 Impact of globalization on entrepreneurship
Source: Author created, according to World Bank Data (2019a).

2.1.2.2 Business Environment

A favorable business environment is one of the basic prerequisites for the development of entrepreneurship, reduction of unemployment and social insecurity and also for the growth of foreign and domestic investments. Namely, the causality between entrepreneurial initiatives and new employment was confirmed (Radović-Marković, 2015).

Among the indicators of the business environment, 11 key ones are listed in Table 2.2.

Here we will analyze the ranking of countries according to four basic indicators (Table 2.3).

Based on the first indicator relating to the establishment of companies, Montenegro is at the bottom of the rankings (90th), among other countries in the region. Only Bosnia and Herzegovina (183rd) is ranked lower than it. On the other hand, Northern Macedonia (47th), Serbia (48th) and Albania (50th) are roughly ranked.

The indicators for the fourth indicator related to property registration indicate that in this case too, Northern Macedonia was best ranked in the list of

Table 2.2 Business environment indicators

Indicators
Establishment of companies
Obtaining building permits
Electrical connection
Property registration
Getting credit
Protection of minority investors
Tax payments
Trading across the border
Time and cost to export
Solving illiquidity
Labor market regulation

Source: World Bank (2019a).

Table 2.3 Ranking of countries by four basic indicators, 2019

	Buss. Esta. Rank		Build. Perm. Rank		Elec. Con. Rank		Prop. Reg.		Rank
Serbia	92.59	48	84.42	11	70.01	10	72.6		55
N. Macedonia	92.08	47	83.38	13	81.43	57	74.5		46
Montenegro	86.65	90	70.88	75	59.19	134	65.78		76
B&H	59.57	183	52.22	167	60.26	130	61.99		99
Albania	91.58	50	57	151	57.71	140	62.08		98

Source: World Bank (2019b).

observed countries (46th place). Bosnia and Herzegovina and Albania ranked worst (99th and 98th, respectively), whereas Montenegro ranked 76th.

2.1.3 Conclusion

The great efforts of the government of Montenegro and other countries in the region have contributed to the progress made by these countries in improving the business environment. Northern Macedonia is ranked best (10th place) by all indicators according to the Doing Business List for 2019, retaining its 2017 position. The reason for this high positioning can be found in the fact that it has implemented comprehensive reforms in the field of small- and medium-sized enterprises. Not only did the reform aim at reducing the time for business registration, abolishing duties and reducing procedures, but through the establishment of a centralized registry, the entire process was transferred from the domain of judicial power to the domain of administration. Further, it should be emphasized that the high ranking of a country means that the government has created a favorable regulatory environment for the operation of business entities.

In Serbia, there has also been a significant improvement in the business environment in the last few years—that is, from 2014 to 2017, by 44 points in rank (Radović-Marković and Tomaš, 2019). However, according to the latest Doing Business List for 2019, Serbia has fallen by one position compared to 2017, whereas Montenegro has progressed by one place and reached the 50th position (Radović-Marković and Tomaš, 2019).

Although significant measures have been taken in Montenegro to improve the business environment by adopting new laws in harmony with EU regulations and implementing institutional reforms in the fiscal system and the financial sector, no significant results have been achieved yet, which is reflected in the fact that, according to all the indicators of the business environment, it is among the worst ranked. Namely, on the basis of indicators, the establishment of companies in Montenegro was in 90th place, whereas on the basis of obtaining a building permit and registration of property, it took 76th place. It was even ranked 134th in obtaining electrical connections. The reason why it is so low on the list regarding obtaining an electricity connection is that it is a very slow procedure that takes 152 days (World Bank, 2019a).

According to the latest Doing Business List (2019b), Bosnia and Herzegovina has experienced a setback from 2017 (−8). The reason is that it has deteriorated in position on some indicators, such as business establishment (−9) and obtaining building permits (−7). It had only made progress in obtaining an electricity connection (+3), until it changed its position in property registration in the last two years.

Albania has worsened in ranking over the past two years (−5). Analyzed by indicators, it also declined when starting a business (−4). It has fallen the most in ranking in terms of obtaining building permits (−65). The reason is that the number of days for obtaining these permits has increased to 299 days (World Bank, 2017, 2019b). Only Albania has made progress on this list in terms of obtaining an electricity connection between 2017 and 2019 (+16).

It can be concluded that all the countries in the region have, generally speaking, progressed in the development of small- and medium-sized enterprises. The most significant improvements are noticeable in the area of introduction of e-government services, which contributes to easier registration, licensing and tax filing process. Nonetheless, there is still plenty of room for improvement in the business environment with a view to accelerating the development of entrepreneurship in the region.

Finally, it should be remembered that the global environment, as discussed earlier, significantly changes standard business and entrepreneurial practices. Today, completely new forms of entrepreneurial activities are emerging. Also, some preexisting forms of activity are gaining new content and are of increasing importance. Accordingly, the emphasis in the development of entrepreneurial activities should be placed on the following areas of economic policy action:

(i) SME development in the green economy.
(ii) Development of a gig-based entrepreneurship or entrepreneurship based on internet platforms.
(iii) Encouraging women's entrepreneurship through programs of empowering women.
(iv) Strengthening the innovation capacity of enterprises.
(v) Supporting startups. These companies use innovative technology, disrupt existing business models and have global outlooks.
(vi) Development of modern entrepreneurial skills through creative education.
(vii) Supporting SMEs to improve their productivity (greater use of alternative financial instruments will further enable SMEs to develop by facilitating their access to finance).
(viii) Improvement of general business parameters so that business internationalization can be faster and resources can be transferred more easily across borders.

References

Acs, Z., Szerb, L., Autio, E., and Lloyd, A. (2017). *Global Entrepreneurship Index 2017*. CreateSpace Independent Publishing Platform.
Ayyagari, M., Kunt, A.D., and Maksimovic, V. (2006). *How Important Are Financing Constraints? The Role of Finance in the Business Environment*. Working Paper No. 3820. Washington, DC: World Bank Policy Research.
Beck, T. and Demirgüç-Kunt, A. (2004). *SMEs, Growth, and Poverty*. Viewpoint No. 268. Washington, DC: World Bank.
Bosma, N. and Levie, N. (2020). *Why Policymakers Need to Understand the Motivations of Entrepreneurs*. Global Entrepreneurship Monitor (GEM).
Calantone, R. and Di Benedetto, M. (2000). Performance and time-to-market: Accelerating cycle time with overlapping stages. *IEEE Transactions on Engineering Management*, 47(2): 232–244.
Deo, S. (2013). *Impact of Globalisation on Small Business Enterprises (SBEs)*. 26th Annual SEAANZ Conference Proceedings, 11–12, Sydney.
Desai, S. (2009). *Measuring Entrepreneurship in Developing Countries*. UNU-WIDER, Research Paper No. 2009/10.
European Commission. (2018). *Strategy for the Western Balkans: EU sets out new flagship initiatives and support for the reform-driven region*. Available at:

https://ec.europa.eu/commission/news/strategy-western-balkans-2018-feb-06_en (accessed on 21 March 2020).

Francis, J. and Richard, F. (2017). Customer service quality management in public transport: The case of rail transport in Tanzania. *International Review*, Faculty of Business Economics and Entrepreneurship, Belgrade and Filodiritto, Italy, No. 3–4: 103–118.

GEM. (2019). *Global Entrepreneurship Monitor, 2018/2019 Global Report*.

Gillie, W. (2018). *Important Factors in Globalization*. Available at: https://medium.com/the-looking-glass/3-important-factors-in-globalization-bde240153502 (accessed on 21 March 2020).

Knight, G. (2000). Entrepreneurship and marketing strategy: The SME under globalization. *Journal of International Marketing*, 8(2): 12–32. https://doi.org/10.1509/jimk.8.2.12.19620

Milner, H.V. and Kubota, K. (2005). *Why the Move to Free Trade? Democracy and Trade Policy in the Developing Countries*. Cambridge: Cambridge University Press.

Munro, M. (2003). A primer on accent discrimination in the Canadian context. *TESL Canada Journal*, 20: 38–51. http://doi.org/10.18806/tesl.v20i2.947

Pologeorgis, N. (2019). How globalization affects developed countries. *Investopedia*. Available at: www.investopedia.com/articles/economics/10/globalization-developed-countries.asp (accessed on 15 June 2019).

Radović-Marković, M. (2007a). *Gender and Informal Economy: Case of Africa, Developing, Developed and Transition Countries*. Lagos: ICEA and UNESCO.

Radović-Marković, M. (2007b). *Preduzetništvo: proces i praksa* (272 str.). Beograd: Magnus.

Radović-Marković, M. (2008). Managing the organizational change and culture in the age of globalization. *Journal of Business Economics and Management*, 9(1): 3–11.

Radović-Marković, M. (2015). Causality among dual education, reducing unemployment and entrepreneurial initiatives of youth in the countries of the Western Balkans. In *Лидерство и организационно развитие (Leadership and Organization Development)* (pp. 10–19). Sofia: Св. Климент Охридски. ISBN 978-954-07-3946-5

Radović-Marković, M., Brnjas, Z., and Simović, V. (2019). The impact of globalization on entrepreneurship. *Economic Analysis*, 52(1): 56–68.

Radović-Marković, M. et al. (2014). *Virtual Organisation and Motivational Business Management*. Beograd: Alma Mater Europaea—Evropski Center, Maribor Institute of Economic Sciences.

Radović-Marković, M., Nikitović, Z., Vujičić, S., and Kasumovic, A. (2019). *Globalisation and the Role of International Organisations*. Available at: www.filodiritto.com/sites/default/files/2020-02/pagine_da_proceedings_ebook_cz01_final-2.pdf (accessed on 18 January 2020).

Radović-Marković, M. and Tomaš, R. (2019). *Globalization and Entrepreneurship in Small Countries*. New York: Routledge. ISBN 9780367250751

Schumpeter, J.A. (1934). *The Theory of Economic Development: An Inquiry Into Profits, Capital, Credit, Interest and the Business Cycle*. Harvard Economic Studies, Vol. 46. Cambridge, MA: Harvard College.

Valdez, M.E. and Richardson, S. (2013). Institutional determinants of macro-level entrepreneurship. *Entrepreneurship Theory and Practice*, 37(5): 1149–1175.
Vučeković, M., Radović-Marković, M., and Marković, D. (2020). *Koncept Digitalnog Preduzeća I Njegove Virtualizacije*. Kruševac: Trendovi u poslovanju.
WEF. (2020). *This Is How the Western Balkans Will Become more Innovative*. Available at: www.weforum.org/agenda/2020/02/western-balkans-become-more-innovative/ (accessed on 13 March 2020).
Wennekers, S. and Thurik, R. (1999). Linking entrepreneurship and economic growth. *Small Business Economics*, 13(1): 27–56.
World Bank. (2017). *Doing Business*. Available at: https://www.doingbusiness.org/en/reports/global-reports/doing-business-2017 (accessed on 10 March 2020).
World Bank. (2019a). *Western Balkans Regular Economic Report: Fall 2019*. Available at: www.worldbank.org/en/region/eca/publication/western-balkans-regular-economic-report (accessed on 1 March 2020).
World Bank. (2019b). *Doing Business*. Available at: www.doingbusiness.org/content/dam/doingBusiness/media/Annual-Reports/English/DB2019-report_web-version.pdf (accessed on 15 March 2020).

Part III
Digital Entrepreneurship

3 Enterprise Digital Transformation Toward Network Virtualization

Mirjana Radović-Marković

3.1 The Digital Enterprise Concept and Its Virtualization

3.1.1 Introduction

The concept of digital company refers to intercompany agility, which integrates resources on a global level by using modern information and communication technologies. When it comes to a manufacturing company, as a condition of agile production, a requirement is set for the establishment of fractal distributed production. It enables the company to have a functional integration of business units. In this way, the manufacturing company increases its speed of response to market demands (Calantone and Di Benedetto, 2000). Digital strategy is an integral part of modern business strategy. In order for a company to be agile, one of the conditions is to design a dynamic production strategy. This principle of agility is successfully applied by the most well-known software companies (e.g., Microsoft, IBM and others).

Virtual companies exist on the principles of digital companies, which implement all their documents and communication in digital form according to some of the standards of formation, storage and exchange of documents. The reference model of a digital enterprise is designed according to the principle of architecture of the open connectivity system (open system architecture)[1] (Marković, 2015). Thanks to their heterogeneity, virtual enterprises can achieve overall success in their outdoor activities although their indoor units are highly specialized (Figure 3.1). Internal specialization generally enables large savings and increases efficiency. They differ from traditional organizations in a large number of characteristics, such as (Ivan, Ciurea, and Doinea, 2012):

1. geographically dispersed;
2. continuously adjusting organizational forms;

3. very good virtual management;
4. intensive use of information technology.

3.1.2 Company Networking

The spatial arrangement of the components of a virtual enterprise is an important and necessary condition for the flexibility of the system. The organization that initiates the cooperation determines the most suitable business processes that are complementary to the business skills of different companies. Accordingly, they combine their knowledge and experience, share business costs and jointly appear on the market in a new virtual form of organizational structure. They are mainly characterized by modularity, heterogeneity and spatial and temporal distance. The main goal of the virtual network company is to enable the participating organizations to rapidly develop their work environment, and the satisfaction of common goals is achieved through a number of resources provided by the participating organizations (Martinez, Fouletier, Park, and Favrel, 2001).

The synergistic effect, which is the result of combining all the basic competencies, enables the formation of an organization with a flexible way of meeting the requirements of users. According to Van Aken (Aken, Hop et al., 1998), a virtual organization must have its own identity. If the identity of one organization remains visible in addition to the links with other organizations, it is considered a "loosely connected virtual organization"

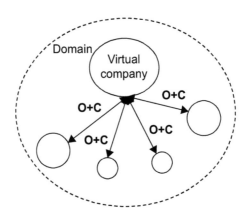

Figure 3.1 Virtual companies

Source: Author.

whereas a "tightly connected virtual organization" looks like one integrated organization with another partner.

Equality of partners in a virtual organization enables the organization to function without hierarchy. The positive effects of such a structure lead to an increase in the efficiency and responsibility of the organization (Bultje and van Wijk, 1998).

3.1.3 Employee Agility

The application of highly flexible production systems alone is not enough to achieve the agility of the company, but agility is much more influenced by the range of characteristics that the engaged workforce must possess, making the main differentiating factors between the agile companies themselves (Figure 3.2).

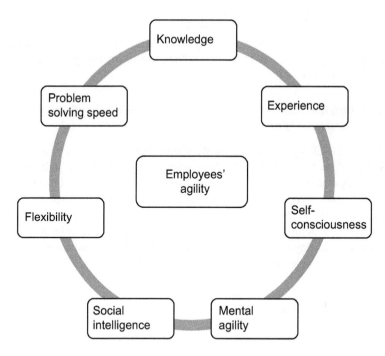

Figure 3.2 Elements of employees' agility

Source: Author.

3.1.3.1 Features of Employee Agility

There are several features of employee agility:

1. *Self-awareness represents a trait where an individual is aware of his/ her strengths and weaknesses.*
2. *Mental agility represents the ability to present problems in a unique and unusual way.*
3. *Social intelligence is reflected in communication skills.*
4. *Flexibility defines the ease of accepting new business challenges.*
5. *Speed of problem solving is the time in which new business challenges are solved.*

By using modern information and communication technology in an agile company, cognitive demand increases, which leads to the need to acquire new knowledge within the educational program of lifelong learning. Modern business requirements have shown that agility cannot be achieved without harmonizing the knowledge and skills of employees (Radović-Marković, 2018a) According to Goldman et al. (1995), in an agile environment, employee skills, knowledge and experience are the main discriminating factors between companies. Therefore, the development of corporate key skills can contribute to sudden changes in the way we do business.

3.1.4 Different Roles of Virtual Teams in the Realization of the Virtual Companies' Functions

Hellriegel (Hellriegel, Slocum, and Woodman, 1997) states that a virtual team is defined as 'a group of individuals who collaborate through information technology on one or more projects, in two or more locations'. The ability to communicate effectively is the basic binder that keeps the team together. Without clear and continuous communication, a team ceases to function, and the common goal of the members of that team in the market can never be achieved.

Duarte and Snyder (2006), in their book, state that virtual teams can be classified into seven basic types:

1. production and development virtual teams;
2. networked virtual teams;
3. parallel virtual teams;
4. production virtual teams;
5. service virtual teams;
6. managing virtual teams;
7. action virtual teams.

What is common to all these types of teams is that team members must communicate and cooperate in order for the task to be done successfully. The following section describes each of them individually:

- A virtual project team or distributed project team includes members in a joint project in which the tasks of team members are not routine, but at the same time, are specific and measurable. This team is additionally characterized as one that has diverse membership, knowledge, systems and jobs and also decision-making capabilities. Team members can change their positions within or outside the project—that is, wherever their expertise is sought. This is often done in order to reduce project costs and to make more efficient use of employees' time and skills throughout the organization (Duarte and Snyder, 2006).
- A networked virtual team consists of individuals who work together to achieve a common goal or task (Duarte and Snyde, 2006). A networked team differs from a project team, in which team affiliation is not always clearly demarcated from the rest of the organization, and the end product is not always clearly defined. Examples of a networked team are often found in consulting firms and organizations based on high technology. The advantages of this type of team are in the fact that it can be formed and disbanded very quickly. These are agile teams, who often use a wide range of experiences, which can help to quickly find creative and innovative solutions to problems.
- Parallel virtual teams perform special tasks that the organization does not want or is not able to perform. The parallel virtual team is similar to the project team in that it also has a special membership, which distinguishes it from the rest of the organization. This makes it clear in the sense of knowing exactly who is part of the team and who is not. Usually members of a virtual parallel team work together on short-term goals such as making recommendations to improve organizational processes or solving specific business problems. Unlike project virtual teams, which are able to make decisions about their goals, parallel virtual teams have more of advisory properties. In a general sense, the formation of virtual parallel teams is becoming a fairly common way of solving problems and making recommendations within a multinational and global organization in global processes and systems that take into account the global perspective (Duarte and Snyder, 2006).
- Unlike virtual project, networked and parallel virtual teams, production virtual teams perform regular and continuous work. Usually such

teams are formed for the purpose of performing a single function, such as accounting, finance, training or research and development.
- Service virtual teams can work in a distributed way through space and time. An example of a virtual service team is customer support to the center, which operates in strategic locations around the world. Their goal is to take advantage of the "follow the sun" strategy (Marković, 2015). This strategy implies that each team works during generally accepted business hours and that the service virtual team at the end of its working day transfers new tasks to the next team that is in a different time zone and is just starting its working day. The desired result is that customers are provided with 24 hours of support seven days a week, and employees are allowed to maintain their typical working hours.
- The managing virtual team of a company can also be detached and remote in space and time. Today, there are many management teams that are scattered across one country or even around the world but still work together on a daily basis. Although they often cross-national boundaries, these teams almost never cross organizational boundaries. Their purpose is to work together to achieve corporate goals and tasks on a business basis as well as to deal with any other topics related to the management of the organization.
- Virtual action teams offer immediate responses, often to crisis situations. They differ from all other types of virtual teams because they are usually formed 'urgently' and only to meet specific needs.

Kauppila, Rajala, and Jyrämä (2011) confirm in their research that the knowledge used by virtual teams is actually knowledge that is known by individual team members but used by the virtual team; the number of individuals who are responsible in the process of knowledge exchange is increased by including them in one wide organizational network. In other words, virtual organizations can improve their internal knowledge exchange by setting up virtual teams, whose explicit task is to share knowledge and persuade other employees to follow the example. It is important to note that the formal knowledge of the whole team gives organizational strength and legitimacy both at the team level and at the individual level.

3.1.5 Importance of Communication in Virtual Enterprises

Communication is essential for the proper functioning of any form of organization, but it is especially important for virtual organizations. Modern communication technologies enable good interaction between its actors who communicate exclusively by electronic means, eliminating the need for physical contact and enabling geographical dispersion of members of the organization. Individuals whose family responsibilities, work obligations or health conditions prevent them from attending meetings can participate and contribute to the discussion through electronic communication.

3.1.5.1 Areas of Electronic Communication

Research into the six areas of electronic communication provides a deeper meaning for understanding the four main aspects of a virtual organization (Monge et al., 1998):

1. strong process dynamics;
2. contractual relations between the entities;
3. bandwidth limits;
4. a structure that is subject to change.

(Marković, 2015)

Most research compares the modalities of electronic communication with spoken language, especially with direct face-to-face communication, despite the fact that electronic communications have many properties similar to the written form of communication. As well as face-to-face communication, electronic communications are interactive. Accordingly, the behavior of electronic communication takes on the characteristics of both types—written document and informal speech (Cooren, Taylor, and Van Emery, 2006).

Successful communication requires that the actors in communication possess the same level of knowledge, which is difficult to achieve without a physical presence. This means that the lack of face-to-face contact in electronic communication can negatively affect the understanding of the message. Research conducted in the direction of understanding electronic communication has come to the conclusion that during the discussion there are several difficulties in understanding the meaning of the information and in managing feedback (Radovic-Markovic, 2011). In addition to the advantages in terms of speed of information exchange electronically and the possibility for participants to communicate at a greater distance, electronic communication has shown that there are certain misconceptions about it—for example, that tasks will not be solved faster if given electronically. However, it has also been proven that in the case of synchronous communication through discussion groups, the lack of visualization did not significantly disrupt the control of the conversation and its comprehensibility (Deans and Kane, 1992). Nevertheless, to resolve some conflict situations and complex activities, as well as to bridge social and cultural differences, visualization is necessary.

3.1.6 The Complexity of Putting Together Project Teams

When determining the type of team, special attention is paid to its structure. The structure itself should be such that the abilities, skills and personality traits of the team members are complementary. Teams with the same or similar professional profiles have not been shown to be effective in practice. In addition, practice has shown that the criteria for electing members, in addition to ability and work experience, should include the personality traits of members. Apart from such traits as energy, perseverance, persistence, tactfulness, cooperation, loyalty to the company and so on, the success of the team depends as much on the ability and expertise of team members. Based on that, a competent virtual team can be assembled that is able to efficiently solve the tasks set before it (Radović-Marković, Tomaš-Miskin, and Marković, 2019).

3.2 Conclusion

For a company to be successful in digital transformation, it must consider the development of many interrelated factors, such as business mobility, infrastructure, data security, customer service and the ability to constantly change and adapt. Therefore, it is not surprising that many organizations, under the threat of digitalization, are trying to introduce new technologies into their business system as soon as possible. Further, companies are increasingly connecting aware of the fact that the new model of business organization requires close cooperation with other organizations. An example is a global organization called the Industrial Internet Consortium, which had 175 members among companies just 18 months after its founding.

Finally, new business requirements have led to the conclusion that agility cannot be achieved without harmonizing the knowledge and skills of employees. New forms of work, new technologies and new requirements set before employees have led to a gradual redefinition of education and to directing individuals and educational institutions in that direction. Training of human resources for certain tasks is conducted primarily by expanding the knowledge and competencies of workers for the best possible response to work tasks (Radović-Marković, 2018b).

Note

1. Open system architecture is a standard that describes a layered hierarchical structure, configuration or model of a communication or distribution system.

References

Aken, J.V., Hop, L. et al. (1998). The virtual organization: A special mode of strong interorganizational cooperation. In M.A. Hitt, J.E. Ricart, I. Costa, and D. Nixon

(Eds.), *Managing Strategically in an Interconnected World.* Chichester: John Wiley & Sons.

Bultje, R. and van Wijk, J. (1998). Taxonomy of virtual organisations. Based on definitions, characteristics and typology. *eJOV Electronic Journal of Organizational Virtualness*, 2(3): 7–21.

Calantone, R. and Di Benedetto, M. (2000). Performance and time-to-market: Accelerating cycle time with overlapping stages. *IEEE Transactions on Engineering Management*, 47(2): 232–244.

Cooren, J., Taylor, R., and Van Emery, E.J. (2006). *Communication as Organizing* (pp. 1–18). Mahwah, NJ: Lawrence Erlbaum.

Deans, P.C. and Kane, M.J. (1992). *International Dimensions of Information Systems and Technology*. Boston: PWS-Kent.

Duarte, D.L. and Snyder, N.T. (2006). *Mastering Virtual Teams: Strategies, Tools, and Techniques That Succeed*, 3rd ed. San Francisco, CA: Jossey-Bass.

Goldman, S.L. et al. (1995). *Agile Competitors and Virtual Organizations: Strategies for Enriching the Customer*. New York: Van Nostrand Reinhold.

Hellriegel, D., Slocum, J.W., and Woodman, R. (1997). *Organizational Behavior*. Seattle: ThriftBooks.

Ivan, I., Ciurea, R., and Doinea, M. (2012). Collaborative virtual organizations in knowledge-based economy. *Informatica Economică*, 16(1): 143–154.

Kauppila, O.P., Rajala, R., and Jyrämä, A. (2011). Knowledge sharing through virtual teams across borders and boundaries. *Management Learning*, 42: 395–418.

Marković, D. (2015). *Integracija virtualnog univerziteta i virtuelnog preduzeća u funkciji povećanja profesionalnih kompetencija zaposlenih*. Doktorska disertacija. Univerzitet Union u Beogradu, Beogradska Bankarska Akademija, Fakultet za bankarstvo, osiguranje i finansije, str. 274.

Martinez, M.T., Fouletier, P., Park, K.H., and Favrel, J. (2001). Virtual enterprise: Organisation, evolution and control. *International Journal of Production Economics*, 74(1–3): 225–238.

Monge, P.R., Fulk, J., Kalman, M., Flanagin, A., Parnassa, C., and Rumsey, S. (1998). Production of collective action in alliance-based interorganizational communication and information systems. *Organization Science*, 9(3): 411–433.

Radović-Marković, M. (2011). *Organisation Behaviour and Culture: Globalization and the Changing Environment of Organizations* (348 pp.). Riga: VDM Verllag Dr Muller.

Radović-Marković, M. (2018a). IT competencies of managers and virtual teams. *Trendovi u poslovanju Business*, 12(2): 1–8.

Radović-Marković, M. (2018b). Modern business environment and entrepreneurship education. In *Digital Transformation: New Challenges* (pp. 148–170). London: Silver and Smith Publishers.

4 Exploring the Synergistic Potential in Virtual University and Virtual Enterprise

Dušan Marković

4.1 Introduction

Social and world economic trends on the one hand, as well as changes in management in practice, communication and organization of work in companies on the other hand, have led to changes in the types of knowledge and ways of acquiring them. In line with this, new educational programs and new ways of learning have emerged. Among the new learning methods, e-learning is undoubtedly the most popular, as it is widely accepted by the faculties that opt for this method of education as well as the students themselves. Namely, almost 90% of all-American colleges offer the possibility of studying online or distance learning, allowing access to their study programs for all potential students wherever they are geographically located. In this way, the market for colleges has expanded, and having in mind simply the number of students studying at virtual colleges (O'Neill, 2012), online education is becoming one of the most profitable businesses in the world. For certain countries, such as Australia or Canada, online education has become one of the most prominent branches of the economy in recent years. In addition to the extremely wide range of study programs offered by these faculties, they can be classified into several categories according to their way of operation:

- Virtual faculties—that is, faculties that offer exclusively online studies.
- Traditional faculties, offering the possibility of combining face-to-face studies with online learning. For example, in a traditional education system, students must choose two or more online courses each school year.
- Faculties that offer special online and traditional face-to-face studies, allowing students to choose between them.

4.1.1 The Concept of Distance Learning

The terms 'e-learning', 'distance learning' and 'distance education' and 'virtual education' are closely related (i.e., education via the internet). The concept of distance learning refers to such learning in which the professor and students are physically distant, and information and communication technology is used to bridge that distance. The key elements of the concept of distance learning and the starting points of its development are technological progress, changes in the educational needs of individuals, active involvement of state authorities at all levels and the establishment of a global educational network.

4.1.2 Prospects for the Development of Education via the Internet

The system of acquiring education via the internet is improving and advancing from year to year in parallel with the development, improvement and application of internet technologies. According to the data (Eurostat, 2020), 87% of citizens in the European Union have had access to the internet in the last year. Although the countries of the Western Balkans lag behind economically in terms of gross national income, they lag slightly behind in terms of internet use (Figure 4.1), which leads to the conclusion that in the last three years, a lot has been invested in communication infrastructure development and IT education. We should not ignore the increase in the living standard of citizens, which enables greater use of information resources.

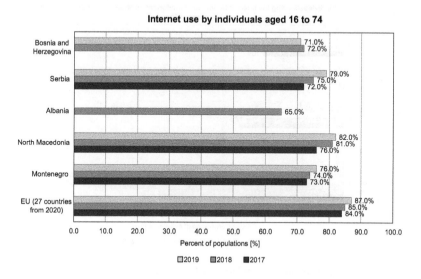

Figure 4.1 Percentage of the population in the European Union using the internet (Eurostat, 2020)

Source: Author created, according to EUROSTAT data.

The increase in the use of the internet by the population (Figure 4.2) refers not only to the use of social networks and games but also for business and educational purposes. Searching for information on education, whether in formal or informal form, via the internet, informs us of the necessity among users for a better form of education as well as their diversity. It is especially necessary to pay attention to the expressed desire for professional training programs. Such training programs are usually implemented outside the university, whereas to increase the quality of their content and knowledge transfer, it would be desirable that they are a part of the offer of the university itself.

The development of wireless internet has greatly contributed to the progress of distance learning, promoting the so-called mobile education, which is not causally related to a particular place of realization. Many software programs, such as the blackboard system and others, have also had a positive impact. With the help of this and other software, students can be in constant contact with their virtual professors. They usually have live lectures or consultations with the professor two or more times a week, and on other days during the semester, the professors ask questions to the students, initiate a discussion, send additional literature, give topics for seminar papers and more. Further, lecturers must offer literature and links to it on the

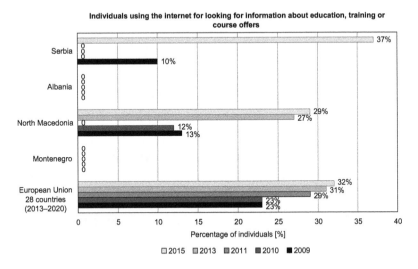

Figure 4.2 Percentage of the population in Europe using the internet to search for training, education information or course offers (Eurostat, 2020)

Source: Author created, according to EUROSTAT data.

internet, and it, of course, must correspond to the students' knowledge and be in the function of the study program. In addition, lecturers or instructors must contribute to creating a greater degree of interactivity when lecturing with students so that they can complete lessons as successfully as possible and get the most out of e-learning programs.

The lecture model created by the lecturer or instructor is like the traditional form of teaching—that is, it integrates the process of planning, implementation and evaluation of the curriculum. Key elements for the development of educational material for online studies are identified within the institution itself, such as its implementation at the technological level and the application of pedagogical methods.

In order to form a successful online course, it is necessary to take into account the opinions of students and their experience gained by attending one of the similar programs; the opinion of professors and their plans, programs and strategies; and put all collected data in the context of lectures/learning.

Despite some differences among experts in terms of how to prepare online programs, almost all of them emphasize the following elements of preparation (Radović, 2010):

- Necessary analyses that should examine the costs, the need to form an online course and the similarities between online programs and traditional courses.
- Identifying the profile of students. It should consider their age, gender, culture, prior knowledge, previous experience with distance learning, computer skills, goals, motivation and so on.

Providing institutional support for distance learning initiatives, such as:

- Vision and mission of the institution
- The lecturers' experience
- Training provided for lecturers and instructors
- Technological infrastructure

Determining pedagogical models and making the choice of the most adequate model should meet the requirements of the program as well as the requirements of students. According to the author (Radović, 2007), this means determining the learning model, learning goals, learning interactivity, development strategy and making choices among numerous web tools (email, chat software, disk board) as well as WebCT software packages or the blackboard system.

The formation of e-learning programs includes making two separate types of decisions: on the one hand, they involve making basic decisions based on learning theory and placing them in a pedagogical framework, whereas on the other hand, they include a whole range of pragmatic decisions, such as those that relate cost and efficiency or quality and safety. In addition to these, other decisions should be included that must fit into the pedagogical framework of learning, forming a very wide range of principles through which the theory of learning is applied in the teaching process and in teaching practice.

Depending on whether students opt for a certified or noncertified program, they may or may not receive a diploma at the end of the completed study program. However, what is most important for many students when obtaining a degree at virtual faculties is the fact that the diploma usually does not state how they studied—that is, whether they studied online or face-to-face'. The reason for this is that these two ways of studying are completely equalized in the world, and no distinction is made between them in employment. The goal of this and other modalities of education and training is to provide students and adults with the knowledge, skills and competencies they need and is sufficient for them to be able to perform one or more related tasks.

In recent years, it is almost difficult to find any faculty in the world who does not offer some form of education with the help of modern technologies and the internet. In this respect, American faculties have advanced the most, who already have a tradition of more than 10 years in this area.

4.2 Creative Education as a Strategic Form of Education

The application of new information and communication technologies has led to rapid changes in the learning environment as well as to the expectation of expertise and a certain level of knowledge of graduates. Based on these expectations, skills in the 21st century are called flexible ability to share work experience. Namely, the needs of professional knowledge pose new challenges for learning and teaching in the educational context. Facing these challenges requires that educational strategies establish new goals for education by placing greater emphasis on individuality.

According to the Millennium Development Goals, education and training are considered a privileged social environment for improving personal individuality and understanding the specific personalities of other people. Accordingly, researchers have begun to focus on the analysis of educational goals and their integration into the process of human development, which includes not only cognitive development but also integration and approach to overall development through a multidimensional formation of human personality and identity. The learning environment should provide conditions for independent learning and support for the human development process. This means that education should help participants to develop analysis skills and a critical spirit with a special focus on researching and evaluating different perspectives. In other words, students receiving a quality education encounter an abundance of intellectual diversity of new knowledge, different perspectives, competitive ideas and alternative truths (Association of American Colleges & Universities, 2019). Thus, when education is successful, it makes us aware that we have the power to change our current situation in a creative way. This means that we become masters of our destinies, not victims of repressive forces.

4.2.1 Encouraging Individuality in Education

Academic freedom implies not only freedom from coercion but also freedom for lecturers and students to work within the scientific community and to develop intellectual and personal qualities for the needs of citizens in life as participants in strengthening the country's economy (Association of American Colleges & Universities, 2019).

A good education system gives students the freedom to recognize their abilities and individual potentials by creating an atmosphere in the classroom in which they promote thinking and reexamine existing conclusions in order to provide students with optimal conditions for personal development (Forte, 2009). In this context, education should encourage students to learn together, ask questions and think creatively about ideas and issues across a range of disciplines. As creative thinkers, they try to imagine and explore alternatives. This approach is necessary for solid academic institutions and the improvement of students' intelligence, including 'soft skills' such as understanding, empathy and communication skills (Dialogue Magazine, 2012). The use of different learning materials allows students with different learning ways to absorb information in the most effective way. Learning fosters multidimensional interactions between students and teachers. For students to learn in their own way, they need many hours to play, explore, overcome boredom, discover their interests and follow them (Gray, 2011). It helps students develop skills of analysis and critical examination with special emphasis on research and evaluation of competitive perspectives. By imaginary engagement in different perspectives, education with greater competence leads to greater personal freedom, but also to greater responsibility (Association of American Colleges & Universities, 2019). However, the student freedom of learning also requires the freedom of teachers to teach (Forte, 2009). In this context, current education systems need to adopt new methods and strategies that can support educational goals and ensure freedom of learning and teaching (Radović and Marković, 2012).

4.2.2 Development of an Education Strategy Based on the Choice of Teaching and Learning

The new education strategy is one that encourages interaction between teachers and students of different needs and ways of learning. This primarily means nurturing creativity, which requires an active way of learning, and thus a new teaching format, where the teacher is a mentor. Creative teachers are ready for changes and welcome new experiences; they are not afraid to go into the unknown. Namely, teachers are the crucial figures in implementing changes, but they need support to understand and accept creativity in their practice. Creative teaching can be defined in two ways: first, as a creative learning, and second, as a study of creativity (Porandokht and Saber, 2013). Creative teaching can be described as a way of enabling teachers to use pedagogical methods to make learning material more interesting, attractive, exciting and effective, and thus attract the interest and attention of students, which is the result of developing a creative approach. The author adds that teachers must strive to better understand their students. In addition, several researchers agree that in creativity, there is always some new, significant and appropriate idea, understanding, information, approach or solution to a problem, given by an individual, group or community. Given the advantage of creativity for society and the individual, one could expect creativity to spread in education (Beghetto, 2005). Creative practice in education should help students work on building their knowledge by defining areas that are especially important to them as well as on strengthening their own sense of individuality. It also includes developing students' personal qualities, including strengthening a sense of responsibility toward themselves and others (Association of American Colleges & Universities, 2019). In other words, the new educational model should be based on individual growth and be able to encourage individuality, whereas flexibility and personality development provide the following characteristics:

- Promotion of achievements.
- Removing barriers to inclusion.
- Creating an environment for learning and teaching that can be adapted to the individual needs of original and creative thinking.
- Intelligent decision making.
- Encouraging young people to learn from experience through multidimensional relations between the concept of the course and the social community.

- Supporting individuals to take ownership of their learning process.
- Improving the relationship between students and teachers, where the teacher is a mentor.
- Acquiring knowledge for problem solving.
- Flexible adaptation to new situations.
- Effective cooperation with others.

4.2.3 The Effects of Educational Strategies to Encourage Individuality

There is an opinion (Radović and Marković, 2012) that the effects of the new strategy in the field of education focus on the individual and the development of his individuality. According to the author, this can be observed from several aspects, such as:

a) The cognitive aspect.
b) The behavioral aspect.
c) The integrated cognitive and behavioral aspect.

Nurturing cognitive abilities helps an individual to develop their intellectual potential and can be achieved through various modern forms of learning, such as:

a) Video games.
b) Computer simulations of real practical situations, which require students to solve a problem or make intelligent decisions.
c) Involvement of participants in research and projects but also engagement in their individual projects.
d) Networking of students within and outside educational communities so that they share knowledge and experience.
e) Participation in continuous discussions on certain topics, which should develop critical thinking, personal attitudes and so on.

The behavioral aspect of conduct should include changes in the student's attitude in relation to the way of studying as well as to develop his ability to master the subject during the learning process. This is primarily related to greater freedom in expressing one's views, which is the basis for encouraging individuality. In addition, students are expected to be independent in learning, whereas the role of teacher/lecturer is reduced to the role of mentor. Teachers should monitor the work of students and guide students. This will help them become independent and self-confident, which is the way that will lead the students to greater autonomy and resourcefulness when they are engaged in the work process.

According to the authors (Radović and Marković, 2012), impressive results cannot be achieved in increasing individuality in the educational process without integrating cognitive and behavioral results. Hence the new learning strategies must respect both aspects. Moreover, it is not only an observation of an important strategy but also its development and change so that it is focused on the needs of both the individual and society. These changes cannot be one-sided but should be viewed from the perspective of the students as well as from the point of view of the teacher. Teachers are also expected to be the perfect promoters of these changes to make the effect of new educational strategies complete. First of all, they will have to accept the role of a mentor and bring themselves to a level where they will be able to lead students in the desired direction—that is, in the direction of strengthening their creativity, originality and logical reasoning. They should provide a relaxed learning atmosphere (stress-free), have good communication with the participants and be always available to them (through modern technologies), showing respect for each student and his cultural, ethnic and gender differences. Teachers should also ensure a balance between structure and freedom of expression and determine the activities that lead students to creative expression (Beghetto, 2005). To achieve these goals, teachers need to become creative and master modern, multidisciplinary knowledge. This, in turn, will be achieved through continuous education but also through joint learning and exchange of experiences in the student-teacher relationship.

4.2.4 The Impact of Information Technologies on Education

High technologies can support creative and innovative learning and stimulate individual potential. The computer as a means of educational technique has proven to be the best tool in the process of individualization because it allows to present educational material in a creative and diverse way. Characteristic of new technologies that facilitate the personalization of learning are different levels of interaction and cooperation (Ferrari, 2009). Namely, the development and implementation of technologies that enable the student to be the most important will bring the need to change pedagogy so that the student is at the center so that he/she becomes the owner of education, which is a necessary quality to encourage creativity (Woods, 2002). It can also support the personal development and maturation of the intellectual. Of course, students can direct their education with their choice of subjects, and a richer choice can be made possible by creating other courses. Increasing the choices also increases the degree of individualization.

High technologies can also improve communication between students and teachers. They provide each student with a greater diversity of learning, improve interactivity between students and teachers, provide personalized learning space, flexible learning outside the classroom walls and allow students to live locally whereas learning globally—through the use of external sources they can access knowledge sources through the internet network.

Many studies compare the modalities of electronic communication with oral speech, especially with direct speech, face-to-face communication, although electronic communications have numerous features of written communication. Like face-to-face communication, electronic communication is interactive. The result is that behavior in electronic communication takes on the characteristics of both the written form and the informal. Although electronic communication has obvious advantages in terms of the speed of information exchange at longer distances, it has still showed some additional doubts. For example, tasks will not be solved faster if they are set electronically.

Appropriate technology platforms require a new approach to learning. They offer many possibilities, such as online learning or a combination of face-to-face learning with online learning. Various online applications can be used to support the teacher and enable him to become innovative in teaching, as well as to help students develop their creative abilities and become creative.

According to one study (Liarokapis and Anderson, 2010), the introduction of a virtual environment in higher education has the potential to bring about positive changes in learning. Namely, the environment for online learning is completely different from the traditional one in the classroom. In other words, online courses require participants to engage in new and different forms of teaching, learning and behavior. Research comparing online learning with the traditional method of face-to-face teaching (Hoben, Neu, and Castle, 2002) examined the effectiveness of online tools, such as discussion boards and chat rooms (Spatariu, Hartley, and Bendixen, 2004), and researchers evaluated the effectiveness of online teaching (Graham et al., 2001) and its value in certain areas of study. Thus, McCombs (2000) gives a list of reasons why he believes that learning online improves the learning process, and among these reasons is that students can learn at any time of the day and at their own pace; they have a lot of information available to them and the ability to track their progress and test their knowledge.

In addition, students who chose e-learning were in an environment where professors responded to their needs and requirements (Radović, 2007).

It can be concluded that the learning process in the classroom can become significantly richer with new technology because students have access to new and different types of information, and they can combine distance learning with traditional face-to-face learning. This combination really opens an unimaginable number of possibilities. Students can do research projects and control experiments in a completely new way, which gives them the freedom to communicate their results and conclusions to the teacher through different media—to the students in their classroom or students around the world.

4.3 Integration of a Virtual University and a Virtual Enterprise

The economy of the Western Balkan countries is not based on knowledge because it does not rely on a well-educated population open to creativity and new ideas. The link between education, research institutes and the commercial sector is very weak, so more investment must be made in the development of innovation policy by providing systemic conditions for the creation, development and implementation of innovations and other measures necessary to encourage the development of overall innovation capacity (Marković, 2015).

Innovation has become one of the most important directions of sustainable development of companies and economic prosperity of the whole society. Enterprises must constantly improve or renew their products and services to remain competitive. Large sums of money are often given to invest in research and development and especially in advertising and marketing of products or services. New investments will not occur if enterprises are not able to recoup their previous costs. That is why it is necessary to work efficiently on innovative projects with clearly defined and realistic expectations. Only such innovative projects lead to greater competence of teachers who work on them, better training of students who would be engaged in projects and their later easier inclusion in the social production segment of society (Ministarstvo, 2012).

The European integration trends point to the need for cooperation between universities and industry to create new values in Europe as a society based on the quality of life and knowledge. This is especially true for industries and companies that do not have their own research units. Without continuing education that provides industry with the innovation of employee knowledge, it is difficult to achieve competitiveness and technological progress.

On the other hand, due to the general crisis in the world, budget allocations for education and science are decreasing, and therefore, the financing of the state of innovative activities will be at an extremely low level. This is especially true in the post-pandemic period, where the world has fallen into a great recession that will probably be worse than the Great Depression of the 1930s. Overcoming this situation can be seen in connecting companies with the university for common interests.

The term 'university industry' implies the establishment of business and information-communication connections of academic institutions (faculties, institutes, university centers) and economic institutions (factories, companies, enterprises) (Marković, 2015). The impact of universities on the industrial increase in the value of products is directly through the cooperation of research and development units of industry and universities and indirectly through the education of experts. Through teaching, the university increases the academic and intellectual values of students to the final value of a graduate expert, following the constant progress of students during their studies. For successful innovation activities, not only regulations and understanding of needs with interpretations of positive practices of developing countries are enough, but above all, the existence of an innovative business climate in which people are motivated for creativity.

4.3.1 Modeling the Integration of a Virtual University and a Virtual Enterprise

Taking into account the considerations given in the previous chapters related to the integration of the virtual university and the virtual enterprise, the contextual level of business cooperation activities was modeled (Figure 4.3). The analysis, definition and modeling of business processes was performed using a top-down method, using IDEF0 notification.

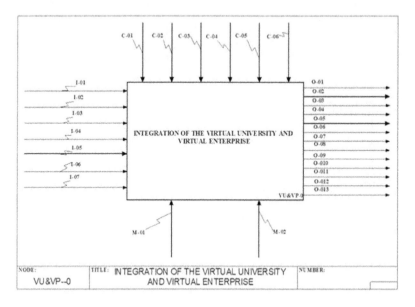

Figure 4.3 The first contextual level of VU and VE integration (Marković, 2015)

Source: Author created, unpublished PhD thesis.

Cooperation within the integration of the virtual university and the virtual enterprise requires a greater number of input parameters that directly affect the work of synergy (VU&VE), which gives richer effects of this cooperation. The demands of individuals and the economy in both education and research and the adjustment of the educational range lead to a better quality of services and educational performance, supporting the paradigm of lifelong learning through formal and informal forms of education (Ministarstvo, 2012) and increasing innovation. The integrated resources of the virtual university and the virtual enterprise are used for the realization of cooperation (Figure 4.3 and Table 4.1, M-01 and M-02), whereas the control mechanisms are the same as for the business of the virtual university (Figure 4.3 and Table 4.1, from C-01 to C-06).

Table 4.1 The input, output and control parameters of the integration of the virtual university and the virtual enterprise

Input		Output		Control		Mechanism	
I-01	User needs	O-01	Curricula	C-01	Accreditation acts	M-01	Virtual university resources
I-02	Reform requirements	O-02	Experts	C-02	Teaching methods	M-02	Virtual enterprise resources
I-03	The needs of society	O-03	Educational performance	C-03	Quality		
I-04	EHEA+ERA	O-04	Research results	C-04	Legislation		
I-05	Business quality methods	O-05	Research performance	C-05	Research methods		
I-06	Budget	O-06	Business performance	C-06	Business methods		
I-07	Contracting jobs	O-07	Business quality methods				
		O-08	Product quality				
		O-09	New products				
		O-010	Service quality				
		O-011	Ned educational programs				
		O-012	Development strategies				
		O-013	Analysis				

The decomposition diagram of the context of the synergy of the virtual university and the virtual enterprise is given in Figure 4.3. There are three basic building activities of the block: namely, the block of the virtual university, the virtual company and the integrated system (VU&VE), which is a jointly formed business space. The integrated system itself is more market oriented, because the virtual university, through such a defined interface, should enable further development of the economy with its needs, both in the form of providing educational services by increasing personal skills of employees and the possibility of product and service development. Due to this orientation, the input parameters into the integral system of synergy are at the same time individual results of the virtual enterprise and the virtual university. The output parameters of the integrated synergy system represent the direct input of the virtual company, whereas the measurement parameters of the control constantly correct the work in the form of system analysis.

Decomposition of the integrated synergy system (VU&VE) (Figure 4.4) represents a modern conceptual solution of the set research task, which in addition to educational activities integrated for the first time professional

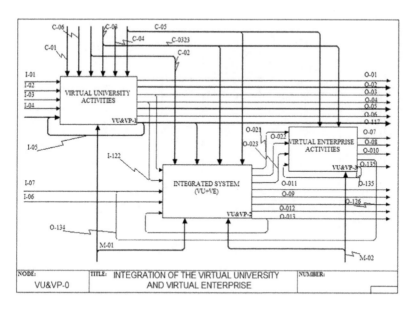

Figure 4.4 Decomposition of the synergy of the virtual university and the virtual enterprise (Marković, 2015)

Source: Author created, unpublished PhD thesis.

training activities (VET) as well as activities that will enable more efficient management of knowledge and trends.

The activities of the expert system (Figure 4.5) are reflected in the acceptance of the requirements of society, companies and European development trends, storing them and forwarding them to other relevant activities. The expert system is supplemented by a knowledge base, and all data realized on other synergy activities are recorded. The interested users who do not belong to the system set requirements for expert opinion in the form of proposals for subjects based on their goals as well as the selection of competent teachers for the implementation of teaching and suggestions of members of the virtual team for the implementation of projects based on keywords.

In addition to storing data on realized synergy activities, the expert system supplies data on other activities in synergy, relying on the knowledge base of the expert system.

The activities of needs and trends analysis (Figure 4.5) performs analyses in all segments of synergy business (VU&VE). Special suggestions and requirements of users are considered, confirming their concept of virtuality, and especially considering the results within business activities. In

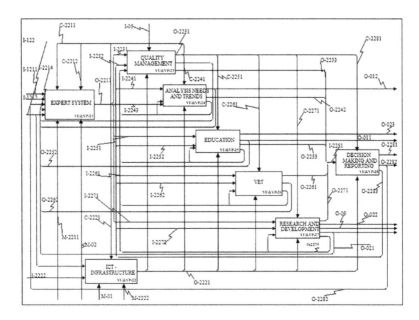

Figure 4.5 Decomposition of the integrated synergy system (VU+VE) (Marković, 2015)

Source: Author created, unpublished PhD thesis.

interaction with the expert part of the needs and trend analysis activities, it forms part of a market-oriented system and represents a form of customer relationship management CRM (client/customer relationship management). The results of this activity are given in the form of documents for further synergy development strategy (VU&VE), which proposes new or modified educational programs and research projects. During the realization of the activity itself, seeking expert opinion is planned, but in the form of a human expert, for questions to which the part of the expert shell cannot offer an adequate answer.

Educational activities implement all necessary activities for the formation of digital educational material that will be used for education during lifelong learning and represents one of the segments of the educational process within the Bologna Declaration (Leuven/Louvain-la-Neuve Communiqué, 2009). In addition to the formation of educational material, it also includes accreditation activities as well as the teaching itself. Because it is about education with work, the most suitable way of teaching is within the electronic teaching platform due to the possibility of different time approaches to learning. During the realization of teaching, the progress of students is monitored through the realization of various preexamination obligations, taking exams or participating in some of the project tasks.

Vocational training activities refer to the activities of acquiring professional skills and are part of the communiqué from 2009 (Leuven/Louvain-la-Neuve Communiqué, 2009), which are realized in the implementation of the training program (vocational education training) at the university as well as in the training laboratories of the industry. The vocational training programs can also be defined as a part of formal education in the form of practical teaching of students, which is provided in current educational plans, but not in the appropriate scope and in the appropriate way. Another form of training plan is defined on the basis of the requirements of the industry to increase the professional competencies of employees in virtual companies. Training programs can be one-time or represent continuous educational forms that are in the domain of the national qualifications framework and are valid for individual countries and are in accordance with European qualifications' frameworks. The realization of training within the laboratory of the industry enables real training of students over the production resources of the industry.

Research and development are activities in which the integrated resources of the virtual university and the virtual enterprise are engaged and where members of the virtual team work. The members of the virtual team are proposed by the expert system, based on the research competencies of the teachers.

Within research and development activities, it includes all activities related to research, development, production and application.

The participation of the virtual team in all phases of the product life cycle increases the quality of the product itself and gains the necessary additional experience.

The virtual university and the virtual company create a virtual connection using information and communication technologies. The very principle of establishing the integration as well as maintaining the connection and the entire IT segment takes place in the part of the activities of the information and communication infrastructure. Monitoring of the IT infrastructure includes measuring the load of the network, database, data and network security as well as servicing the software in the form of installation and settings.

One of the most important activities within the integration of the virtual university and the virtual enterprise is the activity of quality assurance. Each of these activities has its own quality control based on given indicators, whereas a special quality assurance activity has a broader meaning in the form of setting quality standards based on European quality for higher education (EQAR). As part of quality assurance, appropriate quality indicators are set, comparisons are made with current data provided within the data storage, in all areas of activity, and guidelines are given for overcoming possible shortcomings. The data storage is based on analytical data processing; it is a relational database consisting of several fact tables and dimensions. The fact tables contain measurable values of specific business processes, such as students enrolled in individual faculties, elective courses, passed exams, realized projects, published papers of teachers. The dimensions describe the objects that participate in the business, and each dimension joins the business process in which it participates.

Decision-making and reporting activities print reports of all previous activities and forward them to the required users. This activity also includes the decision-making of experts on questions that the expert system itself cannot answer and which are forwarded in digital form. The realization of the expert opinion is monitored within the appropriate activity.

Based on the integration model of the virtual university and the virtual enterprise, a logical database model is modeled, which contains the necessary entities and their relations for the realization of business analysis in synergy (VU&VE). The virtual university allows employees in a virtual enterprise to carry out their educational activities through a lifelong learning (LLL) program as well as the possibility of professional training. The virtual enterprise and the virtual university are working on joint projects that are essential for the existence of the integration of the virtual university and the virtual enterprise.

4.4 Conclusion

Distance learning was conceived in Australia, where education took place by exchanging educational material and solving tasks by mail. With the development of modern information and communication technologies, there has been progress in educational technologies, which now can be used for the presentation of teaching (electronic books, electronic slide presentations, video presentations, simulations, avatar system).

The benefits of distance learning are reflected in the ability to access educational material at any time of the day, from anywhere. Therefore, classrooms no longer have physical walls, but only an electronic environment, which can be accessed by people who meet certain rules. Electronic learning is performed on some electronic platforms with different methodological approaches (blackboard system, Moodle, whiteboard system, blogs), and the student communicates with the subject teacher. This type of teaching is applied to both formal and informal forms of education.

Research in the direction of further development of electronic learning and its perspective has shown that electronic learning will develop moderately in the coming years.

The development of the education strategy based on the application of high technology enables greater creativity in studying in terms of choosing subjects and lecturers. The development of wireless internet and the affordable price of its use enables the reception of signals in a much wider area, and thus greater learning flexibility, as well as the reorganization and adaptation of the platform and educational material to new systems.

At the time of the global COVID-19 pandemic, distance education has become even more important, especially in countries where distance learning is not the predominant way of education. However, in addition to the core activity (i.e., education), the university must perform other activities, such as general university work and research work, which have not been digitized so far. With the virtualization of hitherto nondigitized activities, the university becomes virtual, whereas the IT connection with the virtual enterprise increases the quality of the educational and research process as well as the introduction of new forms of education (VET).

Modeling the process of integration of the virtual university and the virtual enterprise considered all the theoretical rules of existence of complex systems. The processes were modeled from top to bottom so that the system was decomposed to a certain level of complexity. Within the modeling, input parameters, output parameters, control parameters as well as mechanisms for process execution and feedback aimed at system self-optimization are given.

References

Association of American Colleges & Universities. (2019). *Academic Freedom and Educational Responsibility*. Available at: www2.winthrop.edu/acad/AcademicFreedomEducationalResponsibility.pdf (accessed on 11 October 2019).

Beghetto, R.A. (2005). Does assessment kill student creativity? *The Educational Forum*, 69(3): 254–263.

Dialogue Magazine. (2012). *The Freedom to Learn in the Conceptual Age of Schooling*. Available at: www.dialogueonline.ca/freedom-conceptual-age-21st-century-learning-2003/2003/ (accessed on November 2019).

Eurostat. (2020). *Eurostat*. Available at: http://epp.eurostat.ec.europa.eu/portal/page/portal/eurostat/home/ (accessed on 28 May 2020).

Ferrari, A.R.a.P. (2009). *Creativity in Education and Training in the EU Member States: Fostering Creative Learning and Supporting Innovative Teaching*. Seville, Spain: Joint Research Centre Institute for Prospective Technological Studies.

Forte, N. (2009). *Freedom to Teach, Freedom to Learn, the Canadian Centre for Policies and Alternatives, Our Schools-Ourselves*. Montreal: Canadian Teachers' Federation canadienne enseignantes et des enseignants.

Graham, C. et al. (2001). *Seven Principles of Effective Teaching: A Practical Lens for Evaluating Online Courses*. Available at: http://technologysource.org/ (accessed 10 March 2019).

Gray, P. (2011). *Is Real Educational Reform Possible? If So, How?* Available at: www.psychologytoday.com/blog/freedom-learn/201108/is-real-educational-reform

Hoben, G., Neu, B., and Castle, S. (2002). *Assessment of Student Learning in an Educational Administration Online Program*. Available at: www.westga.edu/~distance/ojdla/spring91/chapman91.htm (accessed on 4 March 2019).

Leuven/Louvain-la-Neuve Communiqué. (2009). *Bologna Process European Higher Education Area*. Available at: www.ehea.info/Uploads/Declarations/Leuven_Louvain-la-Neuve_Communiqu%C3%A9_April_2009.pdf (accessed on 10 March 2019).

Liarokapis, F. and Anderson, E. (2010). *Using Augmented Reality as a Medium to Assist Teaching in Higher Education* (pp. 9–16). Norrköping: Eurographics Association.

Marković, D. (2015). *Integration of a Virtual University and a Virtual Company in the Function of Increasing the Competence of Employees*. PhD thesis. Belgrade: Belgrade Banking Academy, Union University.

McCombs, B.L. (2000). *Assessing the Role of Educational Technology in the Teaching and Learning Process: A Learner-centered Perspective*. s.l.: Secretary's Conference on Educational Technology.

Ministarstvo, p.i.n.V.R.S. (2012). *Startegija razvoja obrazovanja u Srbiji do 2020. godine*. Available at: www.mpn.gov.rs/sajt/ (accessed on 15 March 2019).

O'Neill, E. (2012). *3 Reasons Why E-Learning Is Bigger and Better than Ever*. Available at: www.elearners.com/ (accessed on 15 March 2019).

Porandokht, F. and Saber, A. (2013). Creativity in schools. *Procedia—Social and Behavioral Sciences*, 82(3): 719–723.

Radović, M. (2007). Special benefits of e-learning for women: Sample of program entrepreneurship for women. In P. Achakpa (Ed.), *Gender and Informal*

Economy-case of Africa, Developing, Developed and Transition Countries (pp. 156–166). Lagos and Katmandu: ICEA and PRENTICECONSULT with support of UNESCO-a.

Radović, M. (2010). *Tranzicija i nezaposlenost žena starijih od 50 godina* (pp. 261–269). Beograd: Institut ekonomskih nauka, Beogradska bankarska akademija i Savez samostalnih sindikata Srbije.

Radović, M.M.P.F.o.W. and Marković, D.M. (2012). Creative education and new learning as means of encouraging creativity, original thinking and entrepreneurship. *E-Journal of the World Academy of Art & Science Erudition*, 97–114.

Spatariu, A., Hartley, K., and Bendixen, L.D. (2004). Defining and measuring quality in online discussions. *The Journal of Interactive Online Learning*, 2(4): 47–62.

Woods, P. (2002). Teaching and learning in the new millennium. In *Developing Teachers and Teaching Practice* (pp. 73–91). London: Routledge.

Part IV
Changing Nature of Work in the Digital Era

5 The Transformation of Work in a Global Knowledge Economy

Mirjana Radović-Marković

5.1 The Impact of Globalization on New Forms of Labor and Employment

Dynamic changes in the global labor market, the need for innovation and the integration of digital technology in the processes and coordination of economic transactions through digital networks known as 'platforms' have led to changes in business practices and forms of labor. The correlation between different types of changes in an organization indicates changes that have occurred using the electronic functions of the enterprise. This shows, through numerous examples in Europe, a comparative advantage in implementing the new technologies that IT-educated employees have, as they adopt them as new ideas faster than others.

The platforms connect employers with workers from all over the world. The traditional relationship between employer and employee is also changing with the spread of new technology. This has resulted in tectonic structural changes in the global labor supply (Diesel, 2019). Nearly 50% of companies expect that automation will lead to some reduction in the number of full-time employees by 2022 (WEF, 2018). It is also expected to expand jobs and create new employee roles in businesses. First of all, companies are willing to expand the number of employees who will work in specialized jobs, whereas offering them more flexible engagements.

The global market has opened up innumerable opportunities for freelancers, who are outsourced to by firms worldwide. In line with outsourcing procedures, SMEs seek to reduce labor costs when hiring freelancers. Accordingly, 'gig economy' describes an economy based on the accomplishment of tasks and projects, which are increasingly being offered through digital platforms and replacing full-time regular employment (Ciesielski, 2019). Specifically, one of the main differences between this type of employment and the traditional work arrangement is that in this case, workers are paid only for that particular job. Therefore, the jobs are not performed on the basis of permanent employment of people with certain knowledge, but a contract for the provision of a specific service is concluded (Radović-Marković, Brnjas, and Simović, 2019). These companies allow workers quick employment, which can include any type of job. A central argument in the Rockefeller Foundation reports is that platform work offers workers the 'freedom' and 'flexibility' to work whenever and wherever they want (Rockefeller Foundation, 2013).

Jobs offered through 'gig economy' can be classified into two broad categories:

1. For workers with lower incomes and the less educated workforce (craftsmen, suppliers, etc.) who find it difficult to find a job and for whom this work is the main source of income.
2. For workers with higher incomes and the more educated workforce who have other full-time jobs, whereas this type of work generally provides them with additional income.

5.1.1 Types of Platform Companies

Influenced by modern technologies, the way we do business changes and moves from internally focused to externally managed companies. According to the method of running companies, three basic types can be distinguished (Vučeković, Radović-Marković, and Marković, 2020):

1. Innovation platform companies;
2. Transaction platform companies;
3. Integrated platform companies.

- Innovation platform companies. According to some insights, innovation does not occur within companies but is created by interacting with people in the external business environment. Through cooperation with people, new ideas can emerge. Namely, the best combination is if platform companies create the organization, processes and incentives that are necessary to incorporate external ideas. For example, many of the longest-running companies in the list of the world's most innovative companies use platforms to access different data sources, which they then use to build new business models or develop new products and services (Ringel, Laidi, and Djenouri, 2019).
- Transaction platform companies or multilateral platforms. These enable fast transactions between different types of actors. Transaction platforms are usually intermediary platforms. They provide access to third-party products and services on the internet (Yablonsky, 2018).

- Integrated platform companies. These are a set of software services that will facilitate communication between functional entities of the enterprise through computer operating systems, connected to computer and communication networks.

5.2 Changed Manager Role

Managing in a foreign environment where social and cultural norms may be completely different requires special knowledge and skills. The specificity is that managers have limited control over the work and engagement of workers who work in this way compared to traditional firms. Their task is to identify the best projects and to find the most competent workers. Managers can also foster collaboration between traditional and contract workers as well as make strategic plans for engaging contract workers. To get the most out of their employees, they need to connect them to the company's goals. Therefore, performance measurement is about identifying and measuring the skills that will best help an individual grow and thrive and get the company to operate successfully. Continuous feedback (as opposed to the annual performance review) plays an increasingly important role in this system, allowing employees to give and receive feedback much more often.

Of great importance is the fact that employees around the world are not afraid of the future of work—that is, that they are absolutely willing to accept and adapt to changes. It is the responsibility of business managers to recognize this opportunity and be proactive in supporting their employees through the creation of specific action plans. Specifically, managers need to integrate information on company strategy, culture, structure and employee development.

5.3 Strengthening Digital Entrepreneurial Skills in the Western Balkans: Long-Term and Short-Term Goals

Digital skills are becoming an important prerequisite for employment worldwide, but a significant portion of the population still lacks the skills needed to function in a digital business environment. Although young people are often considered 'digital connoisseurs', most of them do not actually have enough of these competencies relevant to starting their own entrepreneurial businesses or filling jobs where there is a need for advanced digital skills.

We do not have enough research investigating the impact of digitalization on the economy and society as a whole. Nonetheless, it is evident that there is a widespread trend in Montenegro and other Western Balkan countries in the demand for digital entrepreneurial competencies. In the absence of the same, in the near future, there could be an increase in structural mismatch in the labor market. Those with the lowest levels of digital skills would be the most affected as well as those who are least willing to upgrade their skills. Accordingly, it is necessary to promote digital entrepreneurial skills and introduce them into education programs through different forms and levels of education. Therefore, the need for policymakers to develop a comprehensive and coherent approach to fostering learning and activating digital entrepreneurial skills among young people, whereas measuring progress toward the goals set, is emphasized. Measurement is important for evaluating whether actions have been successful and whether there are areas in which to achieve these goals that need more attention.

In addition to these, there are many other reasons why digital knowledge must be considered a priority in modern education processes. Accordingly, long-term and short-term goals and their impact at the level of the individual and organization, as well as at the level of economy and society, have been determined.

5.4 Global Gig Economic Index

It is difficult to estimate how many people are employed in the 'gig economy'. Many work on multiple platforms at the same time, yet others do this job whereas maintaining full-time employment at traditional firms. Many private agencies and national statistical institutes in Europe and beyond are conducting pilot surveys to estimate the number of people involved in the platform economy, either as consumers or as service providers. However, often these surveys rely on relatively small samples and usually resort to unconventional sampling techniques, so the results obtained are not sufficiently representative. Thus, estimates of the size of the 'platform economy' vary across studies, and this depends in part on the type and nature of the activity of the platform being measured (Eurofound, 2020). Also, some research shows that 'gigers' include millennials and people close to retirement age (Ciesielski, 2019).

According to the global gig economic index, the US is ranked among the top 10 countries in terms of freelancer revenue, whereas Serbia is ranked as the 10th country by the fastest growing market of the 'gig economy' in the world with 19% growth of the freelancers' total earnings (immediately following Russia) (Figure 5.1).

Asia has become the center of the 'gig economy'. In Asia, Pakistan, the Philippines, India and Bangladesh accounted for the largest amount of operating income. In 2019, they together recorded a 138% increase in revenue compared to 2018. Therefore, it is increasingly heard in professional and scientific circles that by 2030, about half of the world's workforce will be 'gigers' (Harper and Winson, 2018).

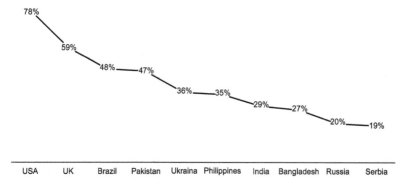

Figure 5.1 Countries ranked by highest revenue from freelancers in the world
Source: Author, according to Diesel (2019).

It is interesting to note that among large countries, a small country like Serbia is present, which belongs to the Western Balkans. Recent research from December 2018 shows that Serbia has 3.52 digital workers per 1,000 population, compared to 1.72 workers in the US (Oxford Internet Institute, 2018).

Although Serbia has over 100,000 freelancers, the law does not recognize them. For some freelancers, this is a major job, and for some it is extra profit. However, there are many who are not registered and do not pay tax, which is detrimental to the Serbian budget.

As the 'gig economy' is growing rapidly, it is imperative to create a policy to optimally exploit this growing phenomenon, which offers numerous opportunities but also has weaknesses. First of all, weaknesses can be reflected in workers' rights. Therefore, there is an increasing call for a legislative response at the European and national level in the field of this type of business (Vučeković et al., 2020). Many Western economies like the UK have already brought a higher level of protection to workers working on 'gig' platforms.

5.4.1 Representation of the Gig Economy in the Western Balkans: Case Study of Montenegro

Given the growth of the 'gig economy', the aim of this research is to determine the interest of the workforce in Montenegro to find employment under this new employment model and work on some of the digital platforms such as Uber, Air Tasker, Delivero and so on. Also, this research should show who the employees are (their education levels, gender, age and the type of jobs they do) as well as the advantages and disadvantages of this type of employment.

Gathering information on workers involved in the 'gig economy' and on the role they play in Montenegro's total revenue can be used to create policies to support this form of employment and help in making legislation.

5.4.1.1 Research Method

The research was conducted via email and a web questionnaire in April and May 2020. We contacted respondents via their social networking profiles (such as Facebook) and professional websites (e.g., LinkedIn). The representative sample consisted of workers who made money in the last year using websites or mobile applications, which connect workers directly with employers.

We selected respondents with the aim to include different experiences of working on the platform, types of jobs completed, education and gender.

> Hypothesis H1. A highly educated workforce sees a great professional opportunity in the 'gig economy'.
> Hypothesis H2. Working through platforms and in the 'gig economy' reduces the brain drain abroad.

5.4.1.2 Key Findings and Discussion

It is believed that freelancing has many advantages such as a better balance of work and private life, the possibility of choosing working hours, high earning potentials, professional advancement and the like. According to our research, conducted on a sample of 150 respondents in Montenegro, certain specifics that are characteristic of the population of Montenegro working in this way are seen.

The analysis performed by the level of education showed that those who had completed university (22.15%) and high school (18.2%) (Figure 5.2) constituted most of the respondents.

Our research results show that the largest number of employees in the gig economy worked in research and education (44%) (Figure 5.3) followed by the provision of programming services (21%) and translation services (20%). These data, to some extent, refute the indigenous opinion that developers work the most in this way.

Transformation of Work 69

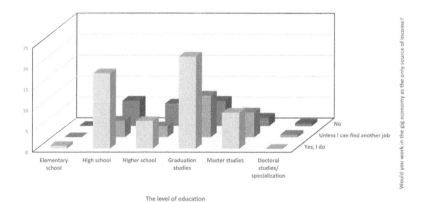

Figure 5.2 Structure of employees in the gig economy by level of education
Source: Author created.

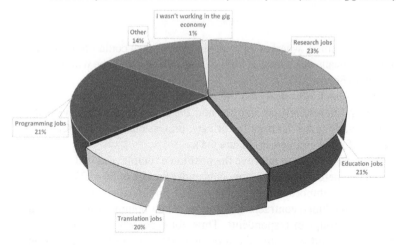

Figure 5.3 The kind of jobs in the gig economy
Source: Author created.

When asked in which domain employees in the gig economy achieve the greatest benefits, the largest number of respondents singled out the professional aspect (20.13%) and only then the financial aspect (18.12%). Further, as many as 56% would accept working in the gig economy as the only

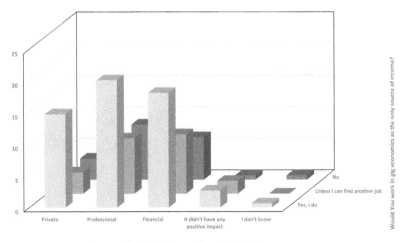

Figure 5.4 In what aspects of life, the gig economy has positively affected
Source: Author created.

source of income, whereas 24% would work if they could not find another job. This means that a total of 80% of respondents would be willing to work this way (Figure 5.4).

The research also analyzed the negative aspects of working in the gig economy. Interestingly, as many as 66.22% of respondents believe that this way of working has no negative private, professional, and financial consequences for the employees (Figure 5.5).

When asked how to improve the position of employees in the economy, the largest percentage of respondents under the age of 30 or 23.6% believe that the most important thing is the trust of employers, on which job security and concluded contracts will depend. Their opinion was not shared by other age groups of respondents. Thus, for those who were in the cohort between 30 and 40 years, the most important is the legal regulation of the position of employees who work in this way (12.8%), given that no country in the region has legislation in this area. The same opinion in the overall structure of respondents is shared by those belonging to the age groups between 40 and 50 years (7.4%) and those older than 50 years (2.0%) (Figure 5.6).

When asked whether working across platforms can reduce brain drain abroad, about 54.4% of respondents answered positively. Among them, 24.83% had completed higher education (Figure 5.7).

Transformation of Work 71

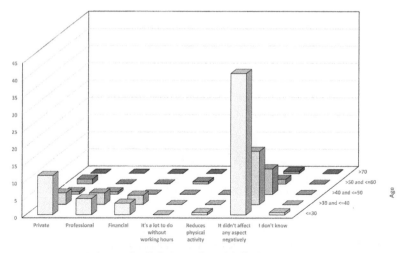

Figure 5.5 In what aspects of life, the gig economy has negatively affected
Source: Author created.

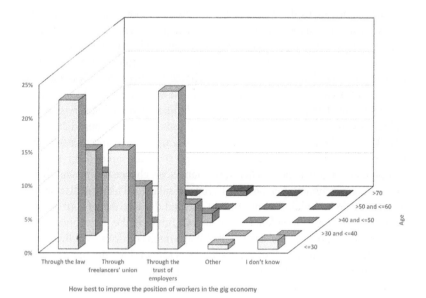

Figure 5.6 How can the position of freelancers be best improved?
Source: Author created.

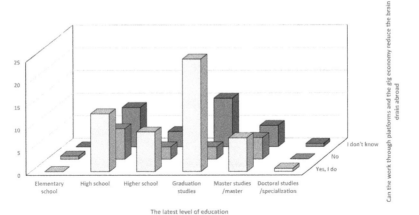

Figure 5.7 How working in a gig economy can reduce brain drain
Source: Author created.

5.5 Conclusion

We are witnesses to the fact that the traditional labor market has significantly changed with the advent of the internet. Until then, the only option for employment was to find a job in a company that would provide the employee with a long-term job, which is located close enough to the employee's residence. This has long been a standard choice not because it was the best option, but because it was the only option. New technologies have opened up employment opportunities outside one's local geographical region and offered a wide range of different well-paid jobs. For Montenegro and non-EU Western Balkan countries, this way of employment has become particularly attractive due to lower wages and high unemployment rates as compared to member states. Therefore, many see a chance to stay and live in the same environment, yet earning European salaries. This was also shown by our research, which was the first of its kind done in the case of Montenegro that 80% of respondents see their own future in the gig economy and singled out the professional aspect (20.13%), which confirms hypothesis H1. In addition, more than half of the respondents (56%) share the same opinion that working across platforms and in the 'gig economy' reduces brain drain abroad, which confirms hypothesis H2. For Montenegro, as the smallest country in the region, this is of particular importance in order to reduce the outflow of the most qualified labor force abroad. The research also showed that as many as 66.22% do not see any negative side of working in this way.

References

Ciesielski, M. (2019). *The Gig Economy Is Becoming Increasingly Global*. Available at: www.obserwatorfinansowy.pl/in-english/new-trends/the-gig-economy-is-becoming-increasingly-global/ (accessed on 12 February 2020).

Diesel, C. (2019). *The Southeast Asian Gig Economy: Capitalism at Its Most Brutal?* Available at: www.aseantoday.com/2019/11/the-gig-economy-capitalism-at-its-most-brutal/ (accessed on 10 February 2020).

Eurofound. (2020). *Telework and ICT-based Mobile Work: Flexible Working in the Digital Age, New Forms of Employment Series*. Luxembourg: Publications Office of the European Union.

Harper, A. and Winson, R. (2018). Deliveroo workers fight for justice-legal ruling will impact on employment rights throughout gig economy. *New Economics Foundation*.

Oxford Internet Institute. (2018). www.oii.ox.ac.uk/blog/tag/gig-economy/ (accessed on 15 February 2020).

Radović-Marković, M., Brnjas, Z., and Simović, V. (2019). The impact of globalization on entrepreneurship. *Economic Analysis*, 52(1): 56–68.

Ringel, M., Laidi, R., and Djenouri, D. (2019). Multiple benefits through smart home energy management solutions: A simulation-based case study of a single-family-house in Algeria and Germany. *Energies*, 12: 1537.

Rockefeller Foundation. (2013). *Health Vulnerabilities of Informal Workers*. Available at: https://assets.rockefellerfoundation.org/app/uploads/20130528214745/Health-Vulnerabilities-of-Informal-Workers.pdf (accessed on 16 January 2020).

Vučeković, M., Radović-Marković, M., and Marković, D. (2020). *The Platform Economy and Flexible Working in the Digital Age*. Konferencija, CANU, Podgorica, June 2020.

WEF. (2018). *The Future of Jobs Report 2018*. Available at: www3.weforum.org/docs/WEF_Future_of_Jobs_2018.pdf (accessed on 10 March 2020).

Yablonsky, S. (2018). *Multi-Sided Platforms (MSPs) and Sharing Strategies in the Digital Economy: Emerging Research and Opportunities*. Hershey, PA: IGI Global.

Part V
Organizational and Entrepreneurial Resilience

6 Resilience and Entrepreneurship

Mirjana Radović-Marković

6.1 Organizational Resilience

Resilience is the theoretical basis and prerequisite for achieving sustainable development (Derissen, Quass, and Baumgartner, 2011). Also, some authors agree that there is a close relationship between resilience and sustainability (Wilson, 2017) and fully equate the notions of resilience and sustainability (Heijman, Hagelaar, and van der Heide, 2007). Others, however, notice subtle differences between the two concepts.

Resistance testing is increasingly taking place both at the level of a national economy and at the level of organizations themselves. Smaller countries are thought to be more exposed to external shocks than larger ones (Briguglio and Vella, 2019). Economic vulnerability is related to one of the following factors:

(a) Openness of small countries because of trade 'shocks' in the external environment. 'This is not a matter of policy choice, because small countries have to export a large part of their production and import a large proportion of products in order to survive' (Briguglio and Vella, 2019, 56);
(b) Dependency on strategic imports such as food and fuel, which also exacerbate the open economies' exposure to 'shocks';
(c) Countries' inclination to natural disasters, leading to 'economic shocks' (Briguglio, 2016).

A country's economy cannot be resilient unless organizations are also resilient to negative impacts. Therefore, these are two sides of the same coin (Radović-Marković, 2017).

In literature, social and organizational resilience are most often investigated and analyzed separately. Nevertheless, in order to improve the resilience of a society, it is important for organizations to make the connection between flexibility and organizational competitiveness as well as invest in resilience. In line with this claim, Radović-Marković (2017), in her research, emphasizes that the resilience of a country's economy cannot be achieved unless organizations are also resilient to negative influences. According to a number of experts, the resilience of a business can be considered as its immune system (Chesley, 2016). In line with this view, understanding how an organization is resilient is crucial to developing strategies to strengthen it.

Research has shown that about half of all businesses experiencing external or internal disasters and shocks that do not have effective recovery plans fail to recover within the next 12 months (Radović-Marković, 2018). In order to survive in the aftermath of a disaster, an organization must carefully pre-plan. However, due to frequent non-planning, businesses, employees and shareholders are exposed to rather unnecessary risks. In addition to insufficient planning, businesses have also shown inefficiency in implementing these plans. 'One of the major obstacles to the successful implementation of a business continuity plan in SMEs is the lack of understanding of the importance of business continuity' (Heng and Wong, 2015). In addition, it is necessary to consider that when integrating businesses into global business flows, current and potential business partners will see resilience as a key criterion for collaboration.

6.1.1 The Modern Concept of Resilience

Resistance or resilience is increasingly a multidisciplinary concept, where economics overlaps with other areas of scientific investigation. A modern understanding of the term 'resilience' entails giving equal importance to risk management and business advancement. This means that organizational resilience is not just about 'surviving' and 'viability' of a business, but comprehensiveness in approaching business success. Therefore, organizational resilience is not a defensive strategy, but a positive dynamic strategy aimed at improving business, which enables business leaders to take risks with security (Allen and Velden, 2005). Specifically, entrepreneurial resilience is a dynamic adjustment process that enables business owners to 'look and move forward' despite the difficult market conditions and destabilizing events they face (Bernard and Barbosa, 2016).

More than many other concepts, resilience or resistance represents adaptability and response to disorders and changes (Imperiale and Vanclai, 2016). How much economies deal with 'shocks' depends on various factors, including the political environment, the depth of economic and financial diversification and especially on the quality of institutions and the economic structure of the economy (Radović-Marković and Tomas, 2019). In addition, countries that had well-developed business plans and a disaster risk action plan, as well as enterprise resilience assessment programs, showed a higher resilience index than those that did not (Radović-Marković, 2018).

The speed of enterprise recovery depends on what is planned in advance in response to potential negative events from the external business environment (Radović-Marković, 2018).

According to FM Global Data for 2019 (Figure 6.1), the Republic of Northern Macedonia is ranked 100th. Compared to 2018, the global resistance index has decreased, and Northern Macedonia has dropped in the ranking by 22 places. The deterioration in resilience is primarily due to a decrease in productivity, an increase in reliance on oil and an increase in the rate of urbanization. After Northern Macedonia, Albania ranked 95th, whereas Serbia and Bosnia and Herzegovina (63rd and 70th, respectively) ranked the best (Table 6.1).

6.2 Resilience and Competitiveness

Competitiveness and resilience are topics considered among the most relevant topics that have animated scientific debates in the field of economic development.

In this analysis, we will look at the degree of correlation between economic resilience (ERI) and global competitiveness (GIC).

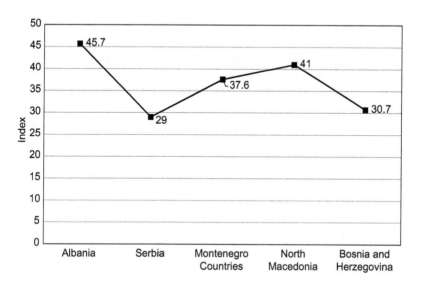

Figure 6.1 Global resilience index

Source: Author, according to FM Global Data (2019) (www.fmglobal.com/resilienceindex).

Table 6.1 Global resilience index for the Western Balkans, 2019

Global resilience index	100	Rank
Serbia	45.7	63
North Macedonia	29	100
Montenegro	37.6	82
Bosnia and Herzegovina	41	70
Albania	30.7	95

Source: Author, according to data FM Global Data (2019) (www.fmglobal.com/resilienceindex).

Economic resilience includes the following indicators (FM Global, 2019):

a) Productivity
b) Political risk
c) Urbanization rate
d) Oil imports

The economic resilience index combines the core drivers of enterprise resilience, highlighting weaknesses and offering guidance to companies, to evaluate and mitigate business risk. In order to ensure the stable operation of organizations, it is important for managers to understand and measure the ability of their businesses to withstand in the event of 'shocks' and respond appropriately to them.

Many small countries are found in high-frequency areas of natural disasters such as typhoons, earthquakes, floods, fires and landslides, which increase environmental vulnerability and affect economic activity. Whereas the world has been facing earthquakes, floods and fires in recent years, 2020 will surely be remembered for COVID-19—a global pandemic that has hit 70% of the world's population.

The pandemic crisis will be the best test for small countries in terms of their economic resilience and the response to such a large 'shock'. Specifically, how this crisis will affect them will be determined by their competitiveness, given their strong reliance on exports.

Considering the relationship between competitiveness and resilience involves understanding how factors combine together can reduce the vulnerability of the territory and the production system and identifying strategies to increase funding for the same factors that are critical to the economic survival and development of the region (Sabatino, 2016).

Our analysis shows that economic resilience and competitiveness are very closely linked (Figure 6.2)—that is, $R2 = 1$.

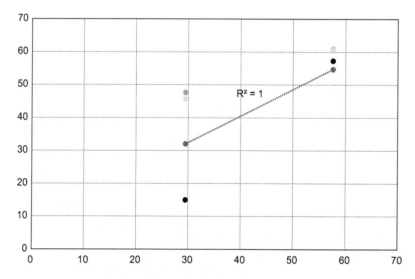

Figure 6.2 The impact of economic resilience on competitiveness

Source: Author, according to data 2019 Resilience Index Annual Report, FM Global (2019).

Further, this analysis showed that the most vulnerable are those companies that are least competitive. Resilience thus emphasizes the importance of flexible enterprises. Creating flexible businesses should be a strategic initiative to change the way a business operates and increase its competitiveness (Sheffi and Rice, 2005; Radović-Marković, 2011). These authors indicate that enterprise resilience can be achieved by reducing vulnerability as well as increasing flexibility, which indicates the ability of the enterprise to return to the 'right path' when disruption occurs. With great resilience, not only will businesses achieve short-term economic success, but they will be able to sustain it in the long term notwithstanding the continued pressure to adapt to changing international competition.

The need for cooperation between Montenegro and other countries in the region on developing regional business risk reduction programs to mitigate catastrophes and greatly reduce costs when they occur should be emphasized.

6.3 Regional Resilience in Times of a Pandemic Crisis: The Case of Montenegro

Resilience implies an adaptation of corporate strategy as well as a solution for organizations that have a high level of threat in all aspects of their work environment. Accordingly, the aim of our research was to determine the how resilient were organizations in Montenegro, especially to the crisis conditions caused by the COVID-19 pandemic.

6.3.1 Research Method

Both qualitative and quantitative methods were used for the study. Descriptive statistics, correlation and regression analysis were used to analyze the data. For qualitative methods, a web-based questionnaire and an in-depth interview were applied. The questionnaire is divided into three parts: (1) biographical information; (2) entrepreneurial resilience; and (3) the business success of the firm and the individual.

There are numerous scales used to measure resilience, the most commonly used are the Connor-Davidson resistance scale (CD-RISC) and the Connor-Davidson 10 resistance scale (CD-RISC 10) (Alstete, 2008). In our study, the CD-RISC-10 scale was used to measure entrepreneurial resilience. The result range is between 0 and 40.

Organizational entrepreneurial success is measured by growth in profitability and sales. A Likert scale ranging from '1, significant decline' to '5, significant increase' was used to measure sales and profitability during the previous year.

Our research is based on testing a two hypotheses set, serving to test the resilience of companies in Montenegro.

> H1. The crisis caused by the COVID-19 virus will not effect the mass closure of companies.
>
> H2. The worst-case scenario is for the company to lose the market in a crisis situation and thus stop working.

Organizational resilience deals with the ability to circumvent disorders through proactive planning, adaptability to a new situation or recovery from shocks and crisis situations. Accordingly, we conducted a research in Montenegro on a sample of 525 respondents (Tables 6.2 and 6.3). The questionnaire consisted of 10 questions relevant to the hypotheses set.

6.3.1.1 Key Findings and Discussion

Given that the research was done at a time when the coronavirus had affected Montenegro and the entire region at the same time, one of the most important questions was 'Do you expect the crisis caused by COVID-19 to affect small businesses in Montenegro?' (Table 6.4 and Figure 6.3).

Based on the answers, it can be concluded that as many as 53.3% of respondents believe that it will not have negative consequences, whereas 23.2% believe that the survival of the companies will be called into question.

Table 6.2 Demographic structure by age

Age		Frequency	Percent	Valid percent	Cumulative percent
Valid	<=30	325	61.91	61.90	61.9
	>30 and <=40	132	25.14	25.10	87.0
	>40 and <=50	44	8.38	8.40	95.4
	>50 and <=60	21	4.00	4.00	99.4
	>60 and <=70	2	0.38	0.40	99.8
	>70	1	0.19	0.20	100.0
	Total	525	100	100	

Source: Author.

Table 6.3 Level of education

Level of education		Frequency	Percent	Valid percent	Cumulative percent
Valid	High school	159	30.30	30.3	30.3
	Higher school	43	8.19	8.2	38.5
	Basic academic studies	256	48.80	48.8	87.3
	Master academic studies	48	9.10	9.1	96.4
	Doctoral studies	19	3.61	3.6	100
	Total	525	100	100	

Source: Author.

We believe that this optimistic scenario of our respondents regarding the impact of the newly emerged crisis is the result of the fact that they gave answers in a situation when it could not be predicted how long it would last and thus what its consequences would be.

One of the questions referred to the worst-case scenario for organizations if they do not have a business continuity plan. Among the offered answers, the respondents felt that the majority of this would be the loss of market position, which itself also means the end of business (39.4%) Figure 6.4).

Immediately followed by the loss of market position, the respondents in Montenegro considered it important for the survival of the company not to lose consumer confidence (33.1%) and then not to damage its reputation (21.5%).

To the question 'How many times did your company stop working until the crisis?', as many as 66.7% of respondents replied that it has never happened, whereas 3.4% were those whose companies stopped working more than three times (Table 6.6).

84 *Mirjana Radović-Marković*

Table 6.4 Do you expect that the COVID-19 crisis will reflect on SMEs?

Do you expect that the COVID-19 crisis will reflect on SMEs?		Frequency	Percent	Valid percent	Cumulative percent
Valid	I will have to temporarily close the company because of the health risk to my employees	83	15.81	15.81	15.81
	I will have to temporarily close my company due to a lack consumers	39	7.43	7.43	23.24
	It will not reflect on the business, because a recovery plan is there	280	53.33	53.33	76.57
	The survival of the firm will be compromised	123	23.43	23.43	100
	Total	525	100	100	

Source: Author.

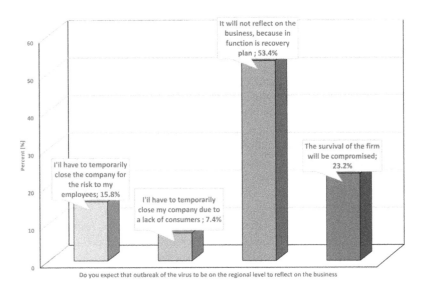

Figure 6.3 Do you expect that the COVID-19 crisis will reflect on SMEs?
Source: Author.

Table 6.5 The worst scenario if the company has no business continuity plan

What is the worst scenario if the company has no business continuity plan?	Frequency	Percent	Valid percent	Cumulative percent
Valid Losing consumer trust	174	33.15	33.2	33.2
Losing the market position	207	39.43	39.4	72.6
The company's reputation is being marred	113	21.52	21.5	94.1
None of these	31	5.90	5.9	100
Total	525	100	100	

Source: Author.

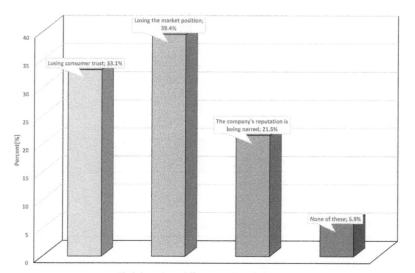

Figure 6.4 Worst scenario

Source: Author.

The answer to this question can be related to the issue of the impact of the COVID-19 crisis on small businesses, where the respondents expressed great optimism about the survival of their firms during and after the crisis, bearing in mind that in the previous period of their operation, they never ceased working. However, in order for the company to be as resistant as possible to external and internal shocks, it is necessary to have a recovery plan in place in case of a crisis. Therefore, one of our questions was to determine the most effective recovery plans. Accordingly, to the question 'How can the most effective recovery strategy be made?', the largest number of respondents thought that this was possible by comparing with similar plans in one or more companies and their implementation in practice (41.1%) (Table 6.7 and Figure 6.5).

A somewhat smaller number were those who gave importance to 'disaster simulation' (33.3%) and based on that made an effective recovery plan.

Our research also shows that a great number of firms (66.7%) have never stopped working during the crisis (Figure 6.6).

Table 6.6 How many times did your company stop working until the crisis?

	You have stopped working in case of impossibility to respond to disasters	Frequency	Percent	Valid percent	Cumulative percent
Valid	Once	83	15.80	15.8	15.8
	More than three times	18	3.43	3.4	19.2
	A couple of times	74	14.10	14.1	33.3
	It's never been	350	66.67	66.7	100
	Total	525	100	100	

Source: Author.

Table 6.7 The most effective recovery strategy

	How can the effectiveness of a recovery plan be determined?	Frequency	Percent	Valid percent	Cumulative percent
Valid	Comparing with one or more other companies to identify the possibilities for recovery plan implementation	216	41.15	41.2	41.2
	Through a simulated disaster	175	33.33	33.3	74.5
	It is not necessary to do anything after creating a plan	54	10.28	10.3	84.8
	I do not know	80	15.24	15.2	100
	Total	525	100	100	

Source: Author.

Resilience and Entrepreneurship 87

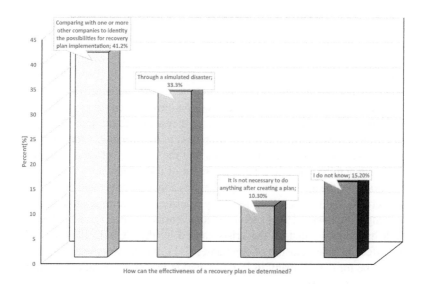

Figure 6.5 How can the effectiveness of a recovery plan be determined?
Source: Author.

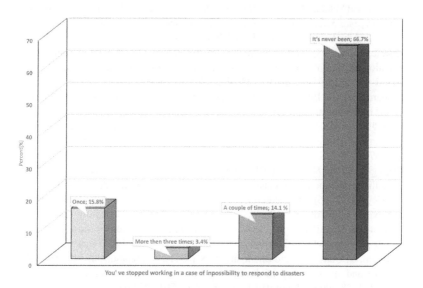

Figure 6.6 Firms who have stopped working in a case of impossibility to respond to disaster (in percent)

Source: Author.

6.4 Conclusion

Based on the obtained research results, several common factors that are crucial to the resilience of the companies have been identified: a) awareness of the situation in which the company found itself, b) the ability to adapt and c) putting into operation an effective recovery plan. The organizations that study the cases and reactions of other companies in crisis situations have received good indicators on how to recover as quickly as possible and keep up their continuity. This research found that there is a correlation between the degree of awareness of the situation, identification and management of vulnerable factors and the capacity of each organization's adaptability. The link between some of these indicators is obvious. For example, the awareness of organizations about the range and impacts of dangerous events affects their recovery priorities, commitment to planning and engaging in simulating such situations. Further, our research confirmed both hypotheses—that is, the crisis caused by the COVID-19 virus will not effect the mass closure of companies (H1), and the worst-case scenario would be for the company to lose the market in a crisis situation and thus cease to operate (H2).

References

Allen, J. and Velden, R. (2005). *The Flexible Professional in the Knowledge Society: Conceptual Framework of the REFLEX*. Maastricht: Research Centre for Education and the Labour Market Maastricht University.

Alstete, J.W. (2008). Measurement benchmarks or "real" benchmarking? An examination of current perspectives. *Benchmarking: An International Journal*, 15(2): 178–186. https://doi.org/10.1108/14635770810864884

Bernard, M.J. and Barbosa, S.D. (2016). Resilience and entrepreneurship: A dynamic and biographical approach to the entrepreneurial act. *Management*, 19: 89–121.

Briguglio, L. (2016). Exposure to external shocks and economic resilience of countries: Evidence from global indicators. *Journal of Economic Studies*, 43(6): 1057–1078.

Briguglio, L. and Vella, M. (2019). The small states of the European Union and the resilience/competitiveness nexus. *Small States & Territories*, 2(1): 55–68.

Chesley, D. (2016). How enterprise resilience can help drive growth in financial services – Enterprise resilience – An important business capability today. *Pwc*, Australia. Available at: www.pwc.com.au/pdf/how-enterprise-resilience-can-help-drive-growth-in-financial-services.pdf (accessed on 15 February 2020).

Derissen, S., Quass, M., and Baumgartner, S. (2011). The relationship between resilience and sustainability of developing societies. *Ecological Economics*, 70(6): 1121–1128. https://doi.org/10.1016/j.ecolecon.2011.01.003

FM Global. (2019). *2019 Resilience Index Annual Report*. Available at: https://fido.nrk.no/4f1683033f5d49fc04861f5a03fa27eb7527ed6e77e81e96f0d6a2fbe5b93dbe/Resilience_Methodology.pdf (accessed on 16 February 2020).

Heijman, W., Hagelaar, G., and van der Heide, M. (2007). *Rural Resilience as a New Development Concept.* EAAE Seminar. Novi Sad: Serbian Association of Agricultural Economists.

Heng, G.M. and Wong, J. (2015). *Business Continuity Management Implementation for Small and Medium Sized Enterprises.* Singapore: BCM Institute Code of Ethics.

Imperiale, A.J. and Vanclay, F. (2016). Using social impact assessment to strengthen community resilience in sustainable rural development in mountain areas. *Mountain Research and Development,* 36(4): 431–442.

Radović-Marković, M. (2011). *Uticaj globalizacije na stvaranje novog modela preduzeća i njegovih konkurentskih prednosti.* Beograd: Radno-pravni savetnik, Poslovni biro.

Radović-Marković, M. (2017). *Podsticanje rezilijentnosti preduzeća u Srbiji, Trendovi u poslovanju Visoka poslovna škola strukovnih studija* (pp. 1–7). Kruševac: "Prof. dr Radomir Bojković".

Radović-Marković, M. (2018). Enhancing resilience of SMEs through insurance. In J. Koćović, B. Boričić, and M. Radović-Marković (уредници), *Serbia in Insurance in the Post-crisis Era* (pp. 155–165). Beograd: Ekonomski fakultet.

Radović-Marković, M. and Tomas, R. (2019). *Globalization and Entrepreneurship in Small Countries.* New York: Routledge.

Sabatino, M. (2016). Competitiveness and resilience of the productive districts in Sicily. The behavior of the Sicilian production areas during the economic crisis. *Contemporary Economics,* 10(3): 233–248.

Sheffi, Y. and Rice Jr, J.B. (2005). A supply chain view of the resilient enterprise. *MIT Sloan Management Review,* 47(1): 41–48.

Wilson, G. (2017). Constructive tensions in resilience research: Critical reflections from a human geography perspective. *The Geographical Journal,* 184(1): 89–99.

7 The Resilient Entrepreneur

Arsen Dragojević

7.1 Entrepreneurial Resilience

Globalization, by its very nature, implies the existence of gradual but far-reaching changes and the establishment of interdependence among various countries and regions. Even increasing global interconnectedness comes with new benefits but also with pitfalls. Whereas this chapter is still being written, the world continues to be in the grasp of the COVID-19 pandemic, and it is exactly this pandemic that has shown in clear light how the benefits of freedom of movement and global interconnectedness besides helping to spread prosperity around the world also help to spread disasters and catastrophes. When one node in global network is affected, this shock reverberates throughout the entire network.

As the pace of global changes is speeding up, the professional community has recognized a need to get deeper insight into the outcomes and consequences that they have on both complex systems and individuals alike. Any form of change is a source of stress, but not everyone handles it the same way, and some deal with it with no problems whereas others are not able to do so. The ability to handle these kinds of challenges is called resilience, but how it is defined exactly? Experts in the field are of consensus that resilience is hard to clearly define. Some authors primarily see this phenomenon as a process (Windle, 2010; Luthar, Cicchetti, and Becker, 2000), whereas others consider resilience to be an amalgamation of multiple individual traits that in unison boost the hardiness of the individual in question. Connor and Davis, authors of the questionnaire that was used in this research, define resilience as 'the personal qualities that enable one to thrive in the face of adversity' (Connor and Davidson, 2003). A multitude of researchers who have covered this topic have pointed out the existence of large number of factors that influence or generate resilience. Some of them point to biological and genetic factors (Tannenbaum and Anisman, 2003; Charney, 2004), some to psychological (Tugade and Fredrickson, 2004) and some to socioeconomic factors (Haskett, Nears, Ward, and McPherson, 2006; Campbell-Sills, Forde, and Stein, 2009).

With all this in mind, we can say that resilience is a complex phenomenon that is influenced by a multitude of factors such as age, gender, culture and that can express itself differently based on current life situation of the individual (Connor and Davidson, 2003). Furthermore, resilience should be distinguished from the idea of anti-fragility because resilience marks the ability to deal with stress (Connor and Davidson, 2003) whereas anti-fragility marks the ability and propensity of a system to thrive and prosper in face of change and adversity (Taleb, 2012). One is focused on nullification of the bad aspects of crisis, while the other is focused on the positive effects of it.

7.1.1 Methodology

Data collection was done through an online questionnaire form, which was delivered via email to a large spectrum of businesses and professional organizations. As interviewees, as part of broader study, were asked to assess the resilience of the organization or company they were a part of; during the recruitment, focus was placed on members with higher levels of education and management. Hence, the sample collected is not representative of the general Montenegrin population, but it represents working population, higher-ranked and more educated employees in particular who are at same time somewhat competent computer users. The sample of 520 valid respondents was collected, covering a large spectrum of fields and industries encompassing such organizations as public institutions, small private businesses and large international companies.

Fieldwork was done during March–May 2020, and its entirety was conducted during the crisis brought on by COVID-19 pandemic. Thus, this was a unique opportunity to measure resilience on a large sample during the period in which the whole population was facing unexpected stressors and challenges. We have measured a self-perceived level of resilience at the exact moment in which the individuals' resilience was tried and tested in real-life situation.

The instrument used in this study was CD-RISC 10 scale of individual resilience. This scale is a shortened version of CD-RISC 25 scale, from which 10 items with the best psychometric characteristics were extracted using the factor analysis (Campbell-Sills et al., 2009). As this was part of a broader study that included multiple scales, it was decided that the shorter (10 items) version should be used to ensure that results are not affected by respondents' fatigue or lack of concentration.

The objectives of this research were:

- to assess the general level of individual resilience in Montenegro;
- to test the influence of demographic characteristics on the level of resilience and to check the results of previous studies that concluded that men, the older population and the better educated have higher resilience scores;
- to study the effects brought on by the pandemic on individual resilience.

During data analysis, descriptive statistics, T-test for independent samples, T-test for single sample, ANOVA and correlation coefficient were used.

7.1.2 Key Findings and Discussion

As no record of the previous use of the CD-RISC 10 scale on a Montenegrin sample was found, the first step was to check the basic metric characteristic of the scale. Testing of Cronbach's alpha yields a result of 0.846, which is satisfactory. Looking into individual questions using corrected-item-total correlation and value of Cronbach's alpha without individual questions, it was concluded that none of questions in the scale warranted removal to improve the characteristics of scale.

To check general level of resilience, total resilience score was calculated by summing up the values of all 10 items. As values of individual questions are in the 0–4 range, the minimal score that someone could get was 0 while the maximum was 40. This new total score was used in all subsequent analyses.

Descriptive statistics showed $m = 28.91$, SD $= 6.61$, and the range of scores was 0–40. What do these results imply? It is apparent that the majority of respondents consider themselves to be highly resilient and on graphical depiction of the frequency-density contour, it can clearly be seen that only a very small number of respondents consider themselves to be of low resilience.

The finding of this study and previous studies on this subject (Campbell-Sills et al., 2009; Antúnez, Navarro, and Adan, 2015) that the majority of respondents consider themselves to be resilient is in line with the idea proposed by George Bonnano, which states that people are far more resilient than is commonly believed (Bonanno, 2004).

As no case of the previous use of the CD-RISC 10 on a Montenegrin sample was found, there was no option to compare data collected during the period of the crisis and during regular periods. In order to find a reference point, relevant literature was reviewed, and the decision was made to compare the data collected with the data collected during the study in which a 10-item questionnaire was created (Campbell-Sills et al., 2009). It is important to point out that an overview of various previous studies has shown that a certain level of cultural and national differences exist for individual resilience. This was to be expected; for instance, resilience level measured in the Spanish sample (Antúnez et al., 2015) has a noticeably lower score than that of the US sample, but it is like the Montenegrin score. To compare results from the US and Montenegrin study, one sample T-test was used. The results indicate that there are significant differences between samples ($t = -7.214$, $p = 0.001$) in which American respondents show significantly higher levels of resilience. One should have in mind that this result is not a definitive answer; it points us to new inquiries. Do Montenegrins have a lower resilience score in general, or was their score affected by current situation? For these questions to be answered, one would need the Montenegrin score during the regular period, which could be used as the baseline.

Due to the fact that data collection lasted over an extended period of time, the effects of progression and development of crisis on levels of resilience could be tested. In order to achieve that, Spearman's coefficient of correlation was calculated for resilience and the time period of data collection was expressed by the number of weeks in a year, but in the end, the correlation was not statistically significant. To eliminate any possibility that there was some relation but not linear in nature, ANOVA was done on the data, where a week of data collection was used as grouping variable, but in this case also, there were no statistically significant results.

Previous studies have shown that certain demographic variables can act as a good predictor of resilience, one among them being gender (Campbell-Sills et al., 2009; Bonanno, Galea, Bucciarelli, and Vlahov, 2007). To test if that is the case in Montenegro, T-test for independent samples was used. The average score for men was 29.25 and for women, 28.62. Indeed, men do have higher scores on average, but checking the value of the T-test shows $t = 1.09$; $p = 0.28$, meaning that existing difference is not statistically significant.

Another claim made in previous studies was that higher levels of education are positively related to resilience (Campbell-Sills et al., 2009). To test this, Spearman's coefficient of correlation was calculated between education level and resilience. The result was $r = 0.059$; $p = 0.17$, meaning that correlation was not significant, and this claim was also not confirmed.

Out of demographic variables that previous studies found to be good predictors of resilience (out of those that we have highlighted), only one left is the age of respondents. Based on research conducted by Bonanno and his colleagues, it is to be expected that older respondents are more resilient to catastrophic events (Bonanno et al., 2007). The correlation of the age of a respondent and resilience is $r = 0.037$; $p = 0.404$. There is no statistically significant link between age and resilience.

7.2 Conclusion

Considering all results, a conclusion can be made that individual resilience is highly spread and present in Montenegro. The majority of respondents consider themselves to be highly resilient, which is another empiric confirmation of Bannon's claim that people are in general more resilient than is believed by the general public (Bonanno, 2004). This result of the Montenegrin sample was lower than the result of the US sample used from a previous study, but it is roughly equal when compared to the Spanish sample.

This leads us to the second important aspect of this study, which is that all results were collected during the COVID-19 crisis. This means that the level of perceived resilience was high during the direct contact of people with a global crisis. It is interesting to note that the level of resilience was stable throughout the crisis, and it didn't fluctuate as the crisis escalated and as more strict prevention measures were being imposed by the government. That said, at the moment, we are not able to establish a direct relation between level of resilience and influence of crisis on it as (as far as we are aware); there is no data on resilience level in Montenegro during the regular period. Is the level of resilience in Montenegro under normal conditions even higher and has it dropped to a lower level in face of the crisis? Is the level of resilience lower in general but now when people are facing a crisis, are they trying to boost their own morale? To these questions and many others, we cannot provide answers without additional study, which can only be done after this crisis has passed.

Not only did the level of resilience prove to be stable across the time but it was also stable across the demographic characteristics. Although some previous studies had implied that certain groups had a higher level of resistance, in this Montenegrin sample, those results were not reproduced. Is it that the perception of resilience is equally distributed in Montenegro? Can it be that direct confrontation with a widespread crisis is somehow bringing scores closer together? Another possibility is that results are mostly affected by sampling method because the sample is specific in that only workers with at least high school diploma and only working population are targeted. This means that there is a possibility that reduction of variance in those characteristics decreases differences across all other demographic indicators. This idea could be tested if the data of previous studies could be analyzed again but only on a subsample of the respondents fitting the sample that was targeted in our study. Once again, we need to point out that all these are just hypothetical ideas and they do not in any way cast a shadow of doubt on previous studies in this area.

With this study, the first insights of resilience levels of Montenegrin society were gained, and additionally insights were gained on the way resilience behaves in a time of crisis. Both are important milestones and foundations for further research in this area. The greatest drawback of this study was the lack of a reference point from Montenegro, which could be used to neutralize possible effects of cultural and socioeconomic differences that tend to affect cross-cultural comparisons. To complete this research, additional follow-up study is needed, which can take place only after the pandemic passes and life goes back to 'normal'. Comparing these two studies would allow us to see exactly how the crisis affected resilience. Further, it is recommended for the next study to cover a representative sample with the unemployed and the lesser educated included, which would shed light on the effects these characteristics have on resilience.

References

Antúnez, J.M., Navarro, J.F., and Adan, A. (2015). Circadian typology is related to resilience and optimism in healthy adults. *Chronobiology International*, 32(4): 524–530. https://doi.org/10.3109/07420528.2015.1008700

Bonanno, G.A. (2004). Loss, trauma, and human resilience: Have we underestimated the human capacity to thrive after extremely aversive events? *American Psychologist*, 59(1): 20–28. https://doi.org/10.1037/0003-066x.59.1.20

Bonanno, G.A., Galea, S., Bucciarelli, A., and Vlahov, D. (2007). What predicts psychological resilience after disaster? The role of demographics, resources, and life stress. *Journal of Consulting and Clinical Psychology*, 75(5): 671–682. https://doi.org/10.1037/0022-006x.75.5.671

Campbell-Sills, L., Forde, D.R., and Stein, M.B. (2009). Demographic and childhood environmental predictors of resilience in a community sample. *Journal of Psychiatric Research*, 43(12): 1007–1012. https://doi.org/10.1016/j.jpsychires.2009.01.013

Charney, D.S. (2004). Psychobiological mechanisms of resilience and vulnerability: Implications for adaptation to extreme stress. *American Journal of Psychiatry*, 161: 195–216.

Connor, K.M. and Davidson, J.R.T. (2003). Development of a new resilience scale: The Connor-Davidson Resilience Scale (CD-RISC). *Depression and Anxiety*, 18(2): 76–82. https://doi.org/10.1002/da.10113

Haskett, M.E., Nears, K., Ward, C.S., and McPherson, A.V. (2006). Diversity in adjustment of maltreated children: Factors associated with resilient functioning. *Clinical Psychology Review*, 26: 796–812.

Luthar, S.S., Cicchetti, D., and Becker, B. (2000). The construct of resilience: A critical evaluation and guidelines for future work. *Child Development*, 71: 543–562.

Taleb, N.N. (2012). *Antifragile: Things that Gain from Disorder*. New York: Random House.

Tannenbaum, B. and Anisman, H. (2003). Impact of chronic intermittent challenges in stressor-susceptible and resilient strains of mice. *Biological Psychiatry*, 53: 292–303.

Tugade, M.M. and Fredrickson, B.L. (2004). Resilient individuals use positive emotions to bounce back from negative emotional experiences. *Journal of Personality and Social Psychology*, 86: 320–333.

Windle, G. (2010). What is resilience? A review and concept analysis. *Reviews in Clinical Gerontology*, 21(2): 152–169. https://doi.org/10.1017/s0959259810000420

Part VI
Economic and Social Impact of Global Ecosystem on Progress in Montenegro

8 Economic Prosperity in the Western Balkans and Montenegro

Mirjana Radović-Marković

8.1 The Economic Prospects and Further Prosperity

The economic growth of Montenegro and the region has raised the standard of living of these countries. However, the standard metric of economic growth over gross domestic product (GDP) only measures the size of total national income generated but does not reflect the well-being of the nation. Nevertheless, economic policymakers often treat GDP or GDP per capita as a comprehensive measure of a nation's development, combining its economic prosperity and social well-being.

Focusing solely on GDP and GDP per capita as a measure of economic development and the well-being of the nation, means the negative effects on society such as climate change and income inequality are ignored (Kapoor and Debroy, 2019). In their study, Kapoor and Debroy (2019) pointed out the need to extend the standard metric of economic growth and to take into account the quality of life and the prosperity of society. This would give a more realistic picture of the economic development of countries at national, regional and global levels. One of the steps in this direction is the Legatum global prosperity index, which measures the economic prosperity of a society.

8.2 Economic Prosperity of Montenegro

Beside the three key factors of economic policy—stable markets, business development and protecting employment—one of the most important is economic prosperity.

Economic prosperity is the key element to quality of life and is also necessary for the nation to be competitive in the world economy (WDM, 2020). In line with this, local economies should move from being production based to ones based on creativity and innovation. However, the prosperity of local economies cannot be ensure without insights into the determinants of their prosperity in a globalized economy. So it is necessary to consider the current COVID-19 crisis, which will lead to a different approach toward the economy and changes at the global level. According to a number of scientists, after this crisis, there must be a strong focus on building more equal, inclusive and sustainable economies and societies that are more resilient in the face of pandemics, climate change and the many other global challenges (UN Report, 2020).

Yet Western Balkan countries will have trouble recovering quickly due to the low resilience of these economies. The Western Balkans are less resilient to cope with the crisis and its consequences (GLOBSEC, 2020). Although the region is one of the poorest in Europe, the crisis will not hit the whole region equally, because there are some countries far better prepared than others. Recent researches conducted by IPSOS and UNICEF have shown that the economy of Montenegro faces a grueling road to recovery from the coronavirus, but the consequences will be successfully overcome with EU assistance (UNICEF, 2020).

8.2.1 The Legatum Global Prosperity Index

The 2019 Legatum Institute improved prosperity index of London measures prosperity in 167 countries around the world. This can be considered from many dimensions. Namely, the index consists of 12 pillars of prosperity (Legatum Institute, 2019):

1) The safety and security pillar measures the extent to which war, conflict, terror and crime have destabilized the security of individuals and left long-lasting consequences.
2) The personal freedom pillar measures progress related to individual freedoms and social tolerance.
3) The governance pillar measures the extent to which governments act effectively and without corruption.
4) The social capital pillar measures the strength of personal and social relationships, institutional trust, social norms and civil society in a country.
5) The investment environment pillar measures the extent to which there is legal certainty for investment.
6) The enterprise conditions pillar measures the degree to which regulations enable businesses to start, compete and grow.
7) The market access and infrastructure pillar measures the quality of infrastructure on which trade depends as well as disruptions in the market for goods and services.
8) The economic quality pillar measures how well a country's economy is equipped to sustainably create wealth and to fully employ its workforce.

9) The living conditions pillar measures quality of life, including material resources, accommodation, basic services and connectivity.
10) The health pillar measures the health of people and the availability of necessary services to maintain good health, including health care facilities, presence of disease, risk factors and mortality rates.
11) The education pillar measures enrolment, outputs and quality at four levels of education (preprimary, primary, secondary and higher education) as well as the skills of the adult population.
12) The natural environment pillar measures aspects of the living environment that have a direct impact on people in their daily lives. It also measures the changes that may affect the prosperity of future generations.

The Legatum index also provides data for a period of 13 years, which is a significantly long time period that can indicate how and in what segments progress can be seen during the same.

Prosperity is much more than material wealth; it encompasses well-being, security, freedom and opportunities. But without an open and competitive economy that stimulates innovation and investment, encourages business and commerce and enables inclusive growth, it is impossible to create lasting social and economic well-being.

Using the Legatum prosperity index, countries around the world can assess their strengths and weaknesses in terms of the level of openness of economies and empowerment of people as well as of an inclusive society. These indicators can serve as a guide for defining development perspectives.

Based on recent data, it can be concluded that global prosperity continues to improve, but the gap between the poorest and most developed countries continues to widen.

According to 2019 data, Montenegro ranked highest in the prosperity index list (50th). It is followed by Serbia, ranked 52nd; Northern Macedonia, 54th; Albania, 65th; and B&H, 70th (Legatum Institute, 2019). Comparing data in 2019 with the one a year earlier, one can see progress in Montenegro by one position and Bosnia and Herzegovina by three positions, whereas Serbia maintained the same ranking. In contrast, Northern Macedonia and Albania fell by one place each in 2019 compared to 2018 (Figure 8.1).

8.2.1.1 The Pillars of Montenegro's Economic Progress

The analysis of the ranking by Legatum's pillar prosperity index shows that Montenegro has made the most progress in terms of investment environment (49th). In terms of market access and infrastructure, it has also made the most progress in the region (50th place) (Figure 8.2).

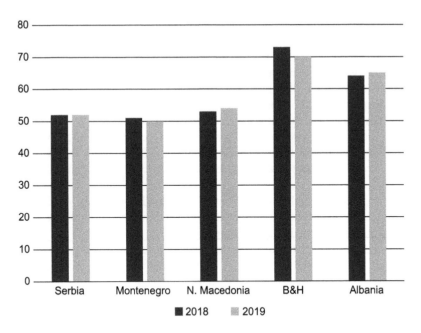

Figure 8.1 Legatum prosperity index, 2018–2019
Source: Author, according to Legatum Institute data.

On the basis of enterprise conditions improvements, which include fast establishment, competitiveness and business development, Montenegro has made the most progress (52nd place). These improvements, combined with the simplification of business administration in starting a business, have made the economy more open.

8.3 Conclusion

In terms of economic quality, Montenegro, at 80th position, came in third place in the region after Serbia and Northern Macedonia. Living, health and education conditions have also improved, although there has been stagnation in the areas of ecology and the environment. Workforce development, education and training all play a critical role in improving economic equality within the country because they create opportunities for local residents to attain good jobs. This, in turn, provides an opportunity to improve the standard of living of citizens across all income categories.

Finally, it can be concluded that in addition to continuous improvement in various aspects of economic development, Montenegro still has room for

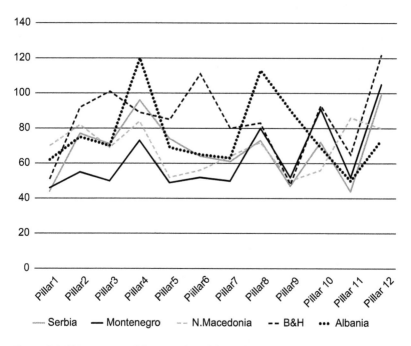

Figure 8.2 The progress of the countries of the region according to Legatum's pillar prosperity index, 2019

Source: Author, according to Legatum Institute Data (2019).

further progress, primarily in the areas of public administration, justice, a functioning market economy and competitiveness.

References

GLOBSEC. (2020). *Coronavirus Makes the Need for a Clear EU Strategy Towards the Western Balkans Ever More Pressing*. Available at: www.globsec.org/2020/04/02/coronavirus-makes-the-need-for-a-clear-eu-strategy-towards-the-western-balkans-ever-more-pressing/ (accessed on 8 April 2020).

Kapoor, A. and Debroy, B. (2019). GDP Is Not a Measure of Human Well-Being. *Harvard Business Review*. Available at: https://hbr.org/2019/10/gdp-is-not-a-measure-of-human-well-being/ (accessed on 18 December 2020).

Legatum Institute. (2019). *Legatum Prosperity Index, 2019*. Available at: www.prosperity.com/feed/topic/regional-analysis (accessed on 18 March 2020).

UN Report. (2020). *Socio-Economic Impacts of COVID-19*. Available at: www.un.org/en/un-coronavirus-communications-team/launch-report-socio-economic-impacts-covid-19 (accessed on 25 April 2020).

UNICEF. (2020). *Public Opinion in Montenegro About Coronavirus, 31 March*. Available at: www.unicef.org/montenegro/sites/unicef.org.montenegro/files/2020-04/IPSOS%26UNICEF%20survey%20March%2031%202020%20ENG.pdf (accessed on 28 April 2020).

WDM. (2020). *Economic Prosperity*. Available at: https://ourwdm.com/communitydesign11121 (accessed on 20 April 2020).

9 Measuring Quality of Life

Borislav Đukanović and Arsen Dragojević

9.1 The Social Prospects: Quality of Life Indicators in Montenegro

9.1.1 Introduction

How best to portray the state of a nation? How to realistically show the true quality of life to which its members are exposed?

In order to answer this question, a dual approach was applied, which should cover the topic from several aspects and make a special 360-degree overview. It includes a view from the inside and the views of the citizens themselves, but also an external view and the assessments of experts from world-renowned institutions. Therefore, the first part is an overview of Montenegrin society and its everyday life through the prism of the original exploratory research conducted directly on the sample of the population in Montenegro. The second segment is a review of a number of quantitative indicators in the form of various indices, which in addition to showing the current situation in Montenegro allow comparing Montenegro with other countries in the region because the full picture of any object can be obtained only if the relevant reference bodies are taken into account.

Our first goal is to describe the everyday life of the inhabitants of Montenegro in the following areas based on empirical data analysis: consumption, family life, free time, cultural practices, value preferences, degree of satisfaction of respondents in different areas of everyday life and time budget.

The second goal is to describe the lifestyles of the strata of Montenegrin society based on the integration of empirical data.

The third goal is to check the acceptability of Bourdieu's theory, primarily based on empirical analysis.

9.1.2 Theoretical Concept of Research

The basic concept of this exploratory research is everyday life. It has historically had different meanings. In the ancient world, everyday life was a sphere of privacy filled by women, children and slaves and which did not concern a free citizen. However, everyday life can also be a training ground for the realization of basic Christian values, through devotion to prayer and religious rituals. With the enlightenment ideals that put the human intellect and its voluntary-moral potentials at the center of social events, everyday life came into the focus of interest. Positivism treats everyday life with certain contempt because the world of everyday life cannot be the subject of positive knowledge. Social turmoil from the end of the 19th and the beginning of the 20th century, and especially Freud's work *The Psychopathology of Everyday Life* indicates that repressed psychic contents, which are related to experiences of everyday life, strongly influence not only psychic processes but also social interactions.

Regardless of the different theoretical approaches, which we cannot deliberate on at this time, some common denominators can be singled out.

First, in the classical difference between the 'abstract' and the 'concrete' in which theoretical reflections on society move, the sociology of everyday life is at the level of the 'concrete'. Instead of their subject of interest being society as a whole, the general features of the culture and sociology of everyday life, they deal with specific events. The sociology of everyday life is in this respect opposed to the categorical thinking characteristic of social theories; whereas the theorists of society operate with initial abstract categories as with empirical facts, sociologists of everyday life try to show how, starting from concrete individuals and concrete relations, those categories arise, transform or disappear.

Second, as a consequence of the focus on the concrete, which is common to different sociological approaches to everyday life, there is a research orientation toward the individual and his everyday practices from which the individual builds his relationship to society (Spasić, 2004).

Third, as sociologists of everyday life are focused on a specific individual and specific events, it is always necessary to situate them in space and time, because only in clear space-time coordinates can we see how individual activities are planned, performed and intertwined.

Fourth, by definition, everyday life includes repetitions and routines, which certainly should not be understood as the common denominator of these routine, ordinary behaviors of an individual, because events and behaviors that are unexpected, unusual and dramatic also take place in it.

Bearing in mind these specifics, which are largely accepted by proponents of various theoretical orientations, we could say that everyday life implies activities, behaviors, more or less goal oriented, with more or less clear meanings of their own and others' activities and behaviors, which individuals practice and repeat every day, in order to adapt or change the narrower or wider social environment in accordance with their wishes, needs, interests and values (Spasić, 2004).

In newer definitions, needs give way to the concept of 'consumption' or 'use' (Spasić, 2004) for which we opted for. This shifts the focus from what people have been given in advance to what people really do. It is important to point out that here the notion of consumption is understood in the broadest sense, not only of material goods but also of symbolic goods. At the same time, this research includes the aspect from the first, older and more dominant point of view in our country, according to which lifestyles are conditioned by class-strata affiliation, which is closer to modern theorists (Bourdieu et al.) Thus, in our hybrid definition of the notion of lifestyle through the class-historical dimension, we have indirectly included the anthropological dimension, as indispensable in its definition. The most important reason for accepting this hybrid notion of lifestyle is the possibility of its more exact operationalization, especially if we consider the transitional character of Montenegrin society. Starting from this compromise solution, we have defined the lifestyle as value patterns of behavior of individuals and groups that influence the choice of consumption of tangible, but primarily of intangible, goods that are specific to those individuals and groups in a given space and time.

Our research design largely supports the basic tenets of the phenomenological sociology of everyday life. First of all, our questionnaire includes numerous variables for a very detailed description of almost all important areas of everyday life, and most questions ask for information on how individuals perceive not only themselves toward others in a particular area of everyday life, but also how they perceive others. Truth be told, in a few areas we have no information on how others perceive respondents. Several other epistemological settings that we encounter in sociologists of everyday life of phenomenological orientation, such as specificity in space and time and goal orientation, support our research design. The phenomenological approach simply imposed itself on us because in several aspects it provided an opportunity for relatively easy operationalization of research issues. However, all these elements, by which our research design supports the phenomenological approach, are in our opinion of relatively less importance than Bourdieu's theoretical heritage, which we consider the most comprehensive and cognitively the most fruitful theory for the sociology of everyday life and lifestyles.

Simply, the results of numerous studies of various aspects of everyday life support Bourdieu's basic theoretical views (Bourdieu, 1990, 1997, 1998; Lamont, 1992; Berstein, 1971; Douglas, 1998; Ostrower, 1998; Allat, 1993).

Therefore, our initial theoretical position is predominantly based on Bourdieu's theoretical opus, which we consider to be the most comprehensive and cognitively the most fruitful theory for the sociology of everyday life and lifestyles. It, as we shall see, is supported by our main research results, as well as numerous other researchers (Bennett et al., 2009; Pakulski and

Waters, 1996; Bourdieu, 2013; Ostrower, 1998). In our research, we have mostly managed to operationalize almost all important theoretical concepts, except those that require more complex and diverse research techniques than the questionnaire.

The central notion of our research is the notion of class in Bourdieu. In Marxist analyses, the prevailing view was about the connection between the way of life and the position of social classes. This attitude has come under fire from critics according to whom social classes have lost their inner coherence and determining role in social life, so that perceptions of the connection of social classes with lifestyles become superfluous (Pakulski and Waters, 1996). Race, nationality, religiousness, gender and age, according to some researchers, have a greater influence than class (Katz-Gerro, 2002). Despite these understandings, Bourdieu's definition of the notion of class is more acceptable to us because, unlike the rigid deterministic understanding of the notion of class in Marxist theoretical heritage, Bourdieu's is much more complex. A class cannot be defined based on the dominant property—for example, economic position of individuals—but Bourdieu constitutes the classes based on grouping and classification of everyday practices. Members of one class are their common habitués, and it is not about class-conscious but class-unconscious processes of forming common habitués. It is important to note that in later papers the term fields gained a similar meaning to the term class.

The next important Bourdieu concept, which we have included in the research, is the notion of practice. The notion of practice has a special analytical value for Bourdieu. Practices, as a sum of individual experiences that are constantly repeated, are the building blocks of everyday life, and everyday life for Bourdieu is a kind of synthesis of the entire subject of sociology. In our research, practices refer to special areas of everyday life—from consumption, family and professional life and free time to cultural practices of value orientations.

Finally, the notion of capital, which consists of all tangible and intangible and cultural and symbolic goods, is operationally included in our research. Consumption is not reduced to the material, but no less to the cultural and symbolic. Under the given conditions, cultural consumption determines the type and scope of other forms of consumption, including material consumption.

The concept of habitus is crucial for Bourdieu's theoretical heritage. Bourdieu understands the habitus as an acquired system of generative schemes that adapt to the special conditions in which they are constituted. As this is a concept that is practically impossible to operationalize by applying research methods such as a questionnaire, we intuitively concluded about it during the analysis of results and complex structural and functional connections of this concept with other mentioned concepts, following Bourdieu's basic theoretical-methodological patterns.

In this case, one important fact should not be lost sight of, which somewhat limits the application of Bourdieu's but also of any other research approach to everyday life: the research of everyday life and lifestyles is connected with open, developed and relatively stable modern societies, which are also quite clearly differentiated at different levels of functioning. All this cannot be said for most societies in Southeast Europe, not even for Montenegro.

These societies in transition are characterized by socioeconomic underdevelopment; undeveloped institutions; unstable, undifferentiated social structure; conflicts on the value and, more broadly, cultural level, with pronounced value confusion and anomic moods of the population. As lifestyles are usually associated with a group or stratum with clear sociological and cultural characteristics, it is quite difficult to expect them to be sociologically profiled in these societies.

9.1.3 Research Method

The SPSS program was used in data processing and analysis. From the statistics, T-test, ANOVA, chi-square test and Mann-Whitney test were used. Pearson's and Spearman's correlation coefficients were used.

When it comes to multivariate methods, Varimax factor analysis and cluster analysis were used.

A sample of 805 respondents for this study has the characteristics of a representative, random and proportional sample. The selection was carried out in three stages. In the first stage, all three regions (North, Central and South) and the capital Podgorica are included. From each region and Podgorica, the number of respondents that was proportional to the total number of respondents for that region—that is, for Podgorica, was randomly selected. In the second step, municipalities were selected. Finally, in the third step, the respondents were randomly selected from the electoral rolls. For the most part, we constructed the questionnaire for this research ourselves. All the previously mentioned areas of everyday life of the respondents are covered with as many as 300 items. The answers to the questions were mostly binary and interval, and less often ordinal and nominal.

Questionnaires were filled out in the field by interviewers, directly interviewing the respondents. We primarily tried to examine the personal beliefs, attitudes and expectations of the respondents, although, as far as our limited resources and time allowed, we tried to determine objective indicators, especially those related to consumption, distribution of family business and time budget.

9.1.4 Key Findings

- *Consumption*

The consumption index was taken over from Lazić and associates (2018), for which we owe them great gratitude. The index is constructed based on total household income during the last year, costs and methods of purchasing clothes and shoes, costs and methods of purchasing personal hygiene products and household cleaners, holiday spending by household members.

Slightly more than a quarter (27.6%) have a high or very high consumption index, and almost half (47.9%) have a low one, whereas about a quarter (24.5%) are positioned around the middle.

Members of the upper classes have a high consumption index; they buy branded clothes, shoes and cosmetics. In the range of these products, they choose the ones of better quality from well-known companies and more often in specialized stores.

They tend to find the most optimal, most rational relationship between price, quality, variety and branding of products, which determines the choice of place of purchase. In addition, they buy selected goods (especially durable consumer goods) in the long run under the most favorable conditions. In contrast, members of the lower social strata in the choice of goods are mostly oriented toward the cheapest goods, where the quality, appearance and variety of goods are secondary to price. Goods are much more often bought by them in Chinese shops (77.5%) and flea markets (69.4%). They indiscriminately buy goods in stores near the house when it comes to food. They also often buy durable consumer goods in private shops, where the prices are relatively the highest and the credit conditions are less favorable. As a rule, they buy necessities, under relatively unfavorable prices and conditions of sale. They buy in places where the variety of goods is low and of poorer quality.

These stratum-class differences and the consumption index are best reflected in food purchases, where the highest stratum spends on average almost two and a half times more money per month than the lowest, with more allocations from the total family budget of lower social strata for food, in relative terms, significantly higher than in higher social strata.

The overall unfavorable economic situation affects the fact that only 57.8% of respondents go on vacation, and of the remaining, only a quarter spend their summers in hotels. The unfavorable economic situation is somewhat corrected by the wives of the respondents with increased savings in the daily distribution of the family budget (49.1%, and all female members in as many as 60% of the cases). They dominate even more strongly in making decisions about the purchase of durable consumer goods. This female "matriarchy" is especially pronounced in the lower and lowest strata (USW and SSW).

We can generally conclude that the structure and type of consumption, the relationship between price, quality and variety of purchased products and services and ways of forming and distributing the family budget are directly and positively related to the level of consumption index, stratification and cultural capital.

In low-income families, women are relatively significantly more burdened with household chores than in those with higher incomes (Giddens, 1991). The results indicate that more diverse and better household appliances in families with higher incomes affects the reduction of women's work in the family. A significant element of negotiating household chores is the income that the partners generate. The researchers point out that these analyses still do not sufficiently take into account not only the technological equipment of the household but also paid assistance, ordering ready meals from restaurants or prepared from the market and other household services, which significantly reduces the share of women's work in higher-income households that are able to pay for all these additional forms of assistance in goods and services (Cohen, 1998; Ruijter, Treas, and Cohen, 2005; Spitze, 1999; Gupta, 2006). These inequalities in the use of different consumer resources can lead to status or even class differences, which have not been sufficiently taken into account in previous research (Gershuny, 2000; Vanek, 1978; Gupta, 2007; Gupta et al., 2009).

Comparing Serbia and Montenegro, it can be concluded that in Serbia, there were more material and aesthetic assumptions and conditions for the development of consumer tastes than in Montenegro (Cvejić et al., 2010). During the prolonged economic crisis of the 1990s, Montenegro became more impoverished, which caused not only an average increase in the consumption index but also the impossibility of developing tastes due to the lack of differences in variety, quality and aesthetic appearance of goods (Tomanović et al., 2006; Bolčić, 1995; Milić, 1994).

The households in Montenegro are thwarted in meeting several important needs. Thus, 15.8% have problems in procuring adequate clothes and shoes, and the same percentage, for financial reasons, avoids visiting friends, and even a third do not go to cafés and restaurants. More than a quarter (28.6%) have problems due to lack of financial resources for elemental housing renovation (e.g., painting) whereas as many as 70% fail for a number of years trying to renovate the living space in accordance with the needs of household members. Satisfying cultural needs is particularly called into question in more than two-fifths of the respondents (40.9%). Procurement of basic household appliances is a problem for nearly a quarter of the respondents (22.9%), and equipping the house with more modern household appliances and modern devices for information and entertainment is a problem for almost two-fifths of the respondents (38.4%). Over a half have a problem getting a car (53.6%), and almost three-fifths (57.8%) cannot think about summer or winter vacation. The fact that more than a third of the households have unmet five or more needs than the mentioned 11 speaks of the low quality of everyday life of Montenegrin households.

Social deprivation is very pronounced, and it is clearly differentiated according to economic, cultural and social capital, which is in line with all the previous results that largely support Bourdieu's theoretical postulates (Bourdieu, 2005, 2013).

As expected, a significant negative correlation was found between the stratum affiliation and the consumption index on the one hand and the deprivation index on the other.[1]

Despite the pronounced deprivation, in the choice of strategy for overcoming financial crises, they rarely use an active approach—that is, attempts to earn extra money (only 12.8%)—but much more often reduce the satisfaction of basic living needs (31.6%) or borrow from banks, or if they do not have the conditions for it, they borrow from friends and relatives. Finally, they resort to selling things from home.

- *Family Life*

Our research also shows that wives do about 50% of the household chores, and female members of the household close to 20%, whereas husbands do about 7%. Therefore, the doers of chores in Montenegrin households are in most cases female members.

The most significant change in Montenegrin society is reflected in the significantly greater participation of husbands in making important family decisions. Wives, on average, in almost 40% of the cases are more likely to make independent decisions on renovation, division and use of housing among household members than husbands, and on average, in a quarter of the cases, more often wives independently decide on the use of annual leave, schooling and extracurricular activities. If we look at the relationship between the participation of spouses in family affairs and decision-making, it is noticed that patriarchal patterns are still maintained because husbands participate almost four times more in decision-making than in performing family affairs and women, only by a fifth; these changes in Montenegrin families can be considered almost revolutionary and key to the transformation of Montenegrin society from a distinctly patriarchal and traditional society until only five or six decades ago to a democratic one.

It is especially interesting that, in addition to husbands, important family decisions are made independently by grandmothers (on average about 6%) and then by daughters (about 2%). It should be emphasized that grandmothers participate many times more in making important family decisions than in family affairs, which indicates that some anachronistic matriarchal patterns are still maintained in Montenegrin families. This is especially the case in the layers of unskilled and semiskilled workers. Although egalitarian tendencies in decision-making are somewhat more pronounced in the higher strata, in which older daughters play a significant role in deciding on children, modern matriarchy is also present in the higher strata, although in the lower strata it is significantly more pronounced and based on pseudo-matriarchal patterns, especially in times of prolonged crises of various types and intensities when the father's mother is the main mobilizer of family members, relatives and friends in resolving these crises.

Depending on the type of family business and decisions, husbands in all social strata are in sharp conflict with their wives over assessments of who is more important in doing family business and especially in family decision-making. They make a much more distorted assessment of the current situation than their wives, in a desire to preserve an idealized image of themselves and strong self-esteem. This image is in sharp contrast to the facts, because even members of the upper classes make significantly less decisions on their own and in making such important and personal decisions such as choosing a profession or professional advancement.

All these divisions result in numerous conflicts in Montenegrin families in 80% of the cases of those who gave answers. Conflicts take place in similar proportions between spouses as well as between parents and their children. However, although conflicts are pronounced in all social strata, their causes have several strata specifics. In the lower social strata, the main source of conflict is the poor financial situation, and in the higher strata, conflicts erupt over the division of family affairs and the upbringing of children. Because of their great ambition and desire to transfer material, social and cultural capital to their children, men belonging to the nomenclature often set high and unrealistic demands on their children that they cannot (or do not) want to meet, probably as an expression of resistance to these excessive demands. When the children of the nomenclature members do not meet high expectations, they resort to bribing children in various ways. The conflicts between spouses in the higher strata are most often about the division of family affairs. As wives in the higher strata are significantly more financially independent and self-conscious, they demand a more balanced division of all household chores.

In the lower social strata, a chronic lack of financial resources and therefore prolonged existential problems result more often in verbal and physical conflicts but also in tendencies that when the crisis reaches its peak, family members look for a way out in stronger internal connections within the family and relatives. We should not lose sight of the fact that these families are more numerous not only because of the larger number of children but also because one or both parents of the spouses live in a joint household with, not so rarely, other close family members.

There is another important area of conflict; families in the higher social strata (especially the stratum of professionals) are relatively often in conflict with relatives and neighbors. In the lower social strata, these conflicts are more frequent with the neighbors. With more pronounced individualistic traits, members of the higher strata (especially experts) come into conflict with patriarchal patterns of relatives, whereas conflicts with neighbors in all, and especially in the higher social strata, appear as opposition to imposed patriarchal moral norms and emphasized demands for collectivist patterns of behavior.

Of the children under the age of 15, 19.3% acquire special knowledge and skills by learning foreign languages, 10.3% by developing artistic inclinations and 18.4% by acquiring special technical knowledge in which computer science occupies a dominant place. It is important to note that in the acquisition of this knowledge, children from the layer of officers and technicians are relatively the most represented. Parents from these strata are trying to increase the cultural capital of their children to increase their chances of ascending the ladder of social stratification. In that respect, this stratum is quite specific in terms of consumption (increased savings) to invest in the social promotion of children.

We specifically examined the correlations between stratum affiliation and satisfaction levels in 10 areas of family life

Table 9.1 shows that there is a positive correlation between the class affiliation of the respondents and satisfaction with most areas of family life.

Table 9.1 Correlations between strata affiliation and satisfaction of respondents in 10 areas of family life

No.	Areas of family life	Ro	Significance
1.	Relationships between family members	−0.125[1]	0.000
2.	Housing situation	−0.118	0.001
3.	Financial situation	−0.240	0.000
4.	Openness, closeness among family members	−0.127	0.000
5.	Connection, solidarity among family members	−0.158	0.000
6.	Harmony of parents in methods of raising children	−0.165	0.000
7.	Common interests of parents	−0.112	0.002
8.	Connection of family members with relatives	0.009	0.780
9.	Connection of family members with friends	0.011	0.775
10.	Health status of family members	0.093	0.008

[1] The negative sign is a consequence of the inverse score, so the negative sign means a positive correlation and vice versa.

Source: Đukanović (2018), unpublished work paper.

Members of the highest stratum (nomenclature) are the most satisfied with the financial situation, connection, solidarity and closeness among family members; harmonization of parents in methods of raising children; relations between family members; housing situation and so on whereas members of the lower classes are the most dissatisfied. No correlations were found between the affiliation of the class and the connection of the family with relatives and friends and the health condition of the family members.

The respondents ranked the most negative occurrences in their families in three ranks.

In the first rank, in the first place was the bad financial situation, in the second place was the housing situation and in third place was the health condition of the family members.

In the second rank, the bad financial situation was again in the first place, the housing situation in the second place and the weak connection of family members with relatives was in third place.

Finally, in the third rank, the first was the poor connection of family members with relatives; second, the health condition of family members; and in third place, the financial situation.

Overall, the respondents were least satisfied with the economic and housing situation, health status of family members and poor relationships with their relatives.

- *Free Time*

Free time best reflects individual personality traits, affinities, tastes and even value preferences. As the person is the least limited in his free time, the freedom and creativity of the person are most pronounced in it. That is why leisure activities are strongly connected with the lifestyles of social strata and individuals in them. We first conclude about this based on the types of activities that individuals like to do—that is, they least want to do in their free time.

Our research shows that watching TV (21%), walking (17.5%) and hanging out with friends (14.5%) are the three most common leisure activities, and in fourth place is sleeping (6.7%). Older and less educated people point out watching television as a favored activity, and younger and more educated people significantly less often. Such a narrow and poor choice of respondents' leisure activities is an important indicator of poor quality of life.

When asked if they dedicate as much time as they would like to their favorite activity, 32.7% said they do not dedicate. Male respondents most often cited the need for additional earnings as the reason, and female respondents the burden of family obligations. Unexpectedly, but in that respect, no statistically significant differences were found, neither in terms of age nor in terms of class and place of residence. When it comes to active or passive leisure time, 39% of respondents actively spend their free time, 30.8% passively and 30.1% mixed. Taken as a whole, women, the younger folk and highly educated people are significantly more likely to actively use their free time.

We were especially interested in the least favorite and the most favorite leisure activities. It is interesting that spouses pointed out as the least favorite the ones that the other spouse considers the most favorite so that the choice spoke much more about marital disagreements about choosing and practicing leisure activities than the result of an autonomous decision. Women significantly more often than men point out sports and recreation, betting and games of chance, fishing and going to matches as the most unpopular activities, and men significantly less often mark these activities as the least favorite. On the other hand, men more often as the least favorite leisure activities stated reading books, beautifying the home, additional education, trips to nature, whereas women list all these activities as the least favorite much less often. It can be concluded that women in their free time focus on activities that contribute much more to spiritual and aesthetic uplift, and they spend their free time in a more diverse and meaningful way than they otherwise do significantly less than men. Maybe that's why they try to make it much better in terms of content and creativity.

Highly educated people are much less likely to consider reading books as the least favorite activity; less educated people are much less likely to consider gambling and fishing as the least favorite activities, and highly educated people significantly more often mark gambling and fishing as the least favorite activities. It is very surprising that no statistically significant differences were found according to strata affiliation.

Finally, what are the three favorite topics to talk about when hanging out with friends in free time? Those are sports (37.99%), children (31.68%) and political events in Montenegro (30.31%). As expected, sports and politics are more important topics for men, and children for women.

With age, interest in children and domestic politics grows significantly, whereas for young people, the dominant topic of conversation is sports. Unexpectedly, the type of topics and their significance are not related to education. Obviously, sports and politics are such universal topics that they cross all educational boundaries. It also looks at the strata boundaries, at least when it comes to politics. Namely, no statistically significant differences were found between the strata when it came to politics as the most favorite topic for conversations with friends in free time. Yet when it came to sports, statistical differences were found; the lower classes significantly more often choose sports as their favorite topic of conversation.

Overall, free time as an important area of everyday life is the least socially profiled. This is expected because it is the least determined by social norms and standards, and in the choice of content and duration, the creativity and freedom of individuals and groups come to the fore. Then it could be expected that in free time, the characteristics of the lifestyle of a certain stratum are well reflected. The results of our research did not confirm this, or at least did not confirm it sufficiently. Namely, the activities that are practiced are quite poor and stereotypical, although, truth be told, the higher strata spend their free time more actively. Although for them that time is of secondary importance due to overwork as well as for members of the lower strata, but not because of the lack of financial resources as with the lower strata, but because of the numerous professional obligations. It is notable that no statistically significant differences were found according to strata affiliation in terms of the least favorite topics and that the choice of leisure activities instead of building areas of common interests and affinities of spouses is a catalyst for marital conflicts, more or less in all social strata.

The choice of the most favorite topics for conversation is also characterized by stereotyping, with politics and sports being the most universal topics, although sports topics are somewhat more popular in the lower strata.

The most significant line of separation of the respondents is neither educational nor stratum related, but gender. It should be especially emphasized that women conceived the content and organization of free time in a richer and more creative way, even at the cost of marital conflicts. We have seen before that women have conquered many areas of deciding on family affairs, even those that were the sovereign privilege of men. This has resulted in increased self-awareness, and it is reflected in autonomous decision-making about the contents and organization of free time.

- *Cultural Practices*

Cultural practice is a complex concept that includes needs, tastes and participation in the consumption of cultural content as well as all the values that influence it. Bourdieu pointed out that the patterns of cultural consumption have a social basis, that tastes and social strata are connected.

According to the traditional understanding of classes, whose constitutive element is a place in the process of production, Bourdieu emphasizes the importance of symbolic meanings that can best be seen in the analysis of cultural practices. In this regard, the notion of habitus has a special significance, which can be interpreted as all internalized patterns of behavior, cognitive schemes and value preferences that arose in the process of socialization. The processes of internalization are class conditioned, especially by the amount and type of cultural capital (Berstein, 1971; Allat, 1993;

Milner and Browitt, 2002; Nemanjić and Spasić, 2006). The main routes of intergenerational capital transfer are educational institutions (Berstein, 1971; Allat, 1993) as well as social networks that access the institutions (Coleman, 1988; Portes, 1998). Of course, the transfer of capital takes place beyond these formal paths. In this regard, personality traits are associated with different amounts of cultural capital (Warde, Martens, and Olsen, 1999). Simply put, a number of studies show the connection between group positions and cultural orientations, when measured by formal indicators of cultural capital—for example, education and professional status (Warde et al., 1999; Warde, 1997; Southerton, 2001; Lamont, 1992)

To collect data on cultural practices in this study, we constructed a Likert five-point interval scale of 25 items. We were aware that with this number of items we could not cover all the emerging aspects of social practices and especially not all social practices that exist in the social space. Nevertheless, we have included the more important ones in the more important aspects.

In this research, we wanted to describe the manifestations of cultural practices in Montenegro and examine the extent to which they are structured. In order to notice some of their structural properties, we applied Varimax factor analysis, which is a good choice when we do not know the scale enough and when we want to single out the factors as clearly as possible, which is appropriate for our case.

The scale is applied for the first time, and that is why we were interested in the extent to which the sample is suitable for factorization. Kaiser-Mezer-Olkin is high and is 0.809, which indicates that the sample is suitable for factorization. When we subjected the 25 items to Varimax factor analysis, five 'pure' interpretable factors emerged that explain 61.518 variances.

The first strongest factor explains 19.744 variances with loads above 0.500. Seven items that describe fans of various film genres, lovers of pop and similar music have the greatest loads on it. We called this first factor the factor of Western pop culture.

On the second factor, which explains 14.412% of the variance, seven items with loads above 0.600 were also singled out, which describe the participation of the respondents in the monitoring of political news programs. We named it the factor of informative-political programs.

The third factor explains 12.18% of the variance. With loads above 0.500, six items describing elite cultural content were singled out on this factor. That is why we have called this factor the factor of elite culture.

The fourth factor explains 9.156% of the variance. The items that describe folk and turbo folk music, as well as reality shows, were singled out. We called it the shun culture factor.

The fifth factor explains 6.01% of the variance. Two items with saturations above 0.800 stood out. In the extended sense of the term, we have called this factor the factor of folk art.

Measuring Quality of Life 119

The population of Montenegro practices various cultural contents, of which only those on the fourth factor can be considered without aesthetic and other values. However, it is a relatively weak factor that carries less than 10% of the variance.

Observed by sociodemographic characteristics, women increasingly accept the elite culture, and men the political-informative programs. The older respondents accept pop and elite culture significantly less, and younger and more educated ones significantly more. The higher strata are much more receptive to elite cultural content, and significantly less to shun culture, whereas there are no significant differences among the strata in terms of monitoring political-informative programs. The contents of the elite culture best separate the higher from the lower strata, and the shun culture best balances these differences. Significant differences are observed in the level of urbanity; residents of the capital Podgorica significantly accept the contents of elite culture than members of other cities in Montenegro.

When it comes to music, which has an important place in Bourdieu's theoretical concept, it is noticed that the lower strata listen to music significantly less than the higher ones. The higher strata listen to pop and classical music significantly more often than the lower ones. Members of the higher strata use much more diverse and numerous ways to acquire and listen to the desired music content than the lower ones, which is partly conditioned by the worse financial situation of the lower strata.

With a two-stage cluster analysis, we tried to examine the grouping of respondents according to the selected factors. A particularly important finding is that the higher social strata are more inclined to reject shun cultural content, and they also show less interest in folk art than the lower ones. However, with few exceptions, social strata have significantly higher frequencies than expected in this cluster. There is a kind of confusion of social strata according to the examined cultural practices.

As the social strata on the fourth cluster are not fundamentally different, cluster analysis showed us that the prevalence of the fourth cluster would be in favor of Peterson's omnivore theory (Peterson, 2005, 1997) rather than Bourdieu's.

With minor exceptions (small private individuals and the social class of the self-employed close to them), they are less interested in informative and political programs. It is obvious that small private individuals and the self-employed do not have a special interest in these programs because they cannot find information that can be practically important to them, and members of other higher classes can (at least indirectly) and are therefore more interested in these programs. In this respect, officials and technicians are closer to the lower social strata because they obviously believe that they cannot use this information more directly for social and cultural promotion, which they want to achieve as soon as possible. In contrast, they prefer elite cultural content, presumably not because they share the same aesthetic values with these strata, but out of a strong need to identify with the higher strata and 'skip' the middle position as a buffer zone between the lower and higher social strata. In support of this is the finding that officials and technicians at the shun culture factor are closer to the lower than the higher social strata.

If we simplify our results on cultural practices, we can say that younger, more educated members of the higher classes and those who come from the capital Podgorica prefer elite and pop culture, and the older and the less educated from the lower classes and other cities of Montenegro prefer more shun culture and folk art. Truth be told, whereas the orientation of the upper strata toward the elites and pop culture is more explicit and clear, the orientation of the lower strata toward shun culture and folk art is not always explicit and clear.

Although fewer findings support omnivore theory (Peterson, 2005, 1997), there is no doubt that the basic results confirm Bourdieu's views (Bourdieu, 1990, 1997, 1998). The first and most significant finding that supports this general conclusion is that the contents of elite culture are significantly more present among members of the upper strata, who inherit higher material, cultural and social capital, which clearly distinguishes them from the lower social strata. Let us remind ourselves that these are younger, more educated people from higher social strata, from Podgorica. Another finding that should also be considered is that more modern cultural content, such as pop music, is again relatively more inherited by members of higher social strata, the younger folk and those with greater cultural capital.

- *Value Preferences of Social Strata*

Miles and his associates view values as motivational forces that influence much behavior of individuals and groups and the creation of social identity (Miles, 2014). Several other researchers believe that values in this context have been neglected (Ignatow, 2009; Joas, 2000).

Bourdieu links values to the convertibility of different types of capital (social, cultural, symbolic and economic) (Lamont, 1992; Bourdieu and Passeron, 1977).

From the point of view of our research, it is especially important that values directly or indirectly influence all activities in everyday life.

To examine the value preferences of the social strata of Montenegro, we constructed the Likert five-point scale, which contains 18 items. During the construction, a dilemma arose as to whether to consistently operationalize value orientations or to adjust the choice of questions to the description of lifestyles of social strata. We opted for a compromise solution.

The internal consistency of the scale is satisfactory (Krombach's alpha = 0.6). Kaiser-Meyer-Olkin = 0.796. It is high and indicates the suitability of the results for factorization (Table 9.2).

The screen plot and the meaningfulness of the factors served as the basis for the choice. As relevant loads, we took those above 0.500, although many were above 0.600 and even 0.700.

The first factor carries 22.224% of the variance. It describes the professional aspirations of family members, followed by health and harmony among them. We called it the family factor.

The second factor carries 14.172% of the variance and is also made up of six items with saturations above 0.600, except for one. It consists of items that describe a secure job, good professional performance and respect for individual rights. We called it the factor of professional success and individual rights.

Whereas other factors are 'clear', for the third, this could not be said because it includes the statement that indicates individualism: 'I think it is important to pursue my personal interests regardless of whether the means used for that are morally acceptable or not'. However, other claims point to authoritarianism, and we called it the factor of predominantly authoritarian value orientations.

The fourth factor carries 5.961% of the variance. Two claims with saturations above 0.700 stand out in it and describe the nationalist value orientation. We called it the factor of nationalism.

Factor scores were compared by sociodemographic characteristics. The only significant difference by gender is in the first factor, where, as expected, females are significantly more represented. According to the other three factors, no statistically significant differences were found. We have already pointed out that women in Montenegro have become much more self-aware and are much more involved in the distribution of family affairs and family decision-making. This is a consequence of the much greater resilience of women. In achieving professional goals, women are no less resilient than men. Having maintained a caring position toward children, they do not lag behind men in the manifestation of authoritarian-nationalist value preferences, which occasionally become important in social promotion.

As expected, young people are more represented on the second factor—the factor of professional success and respect for individual rights, but they do not lag significantly behind the elderly in the manifestation of authoritarian values, although the elderly are hard bearers of authoritarianism and nationalism. One gets the impression that this contradiction in the value orientations of young people is not so much an expression of their essential value determinations, but the need to conform to the desirable social norms of the establishment.

Table 9.2 Matrix of factor saturations of value preference items

	Component			
	1	2	3	4
Quest. 13 It is important to me that children get a permanent and well-paid job.	0.805	0.130	−0.013	−0.053
Quest. 12 It is important to me in life that children achieve the highest and highest quality professional education.	0.767	0.137	−0.048	0.047
Quest. 15 Harmony and understanding among family members are especially important to me.	0.703	0.157	−0.175	0.107
Quest. 14 My health and the health of my family members are especially important to me.	0.687	0.218	−0.206	−0.027
Quest. 16 It is especially important for me to live in a country where there are no upheavals and conflicts so that family members can achieve their life goals.	0.637	−0.088	0.052	0.249
Quest. 18 Love and harmony in marriage are important.	0.604	0.233	−0.032	−0.088
Quest. 5 It is primarily important to me to have a steady and well-paid job.	0.102	0.757	−0.045	0.106
Quest. 6 It's important to do a job that I love and enjoy it.	0.141	0.749	−0.100	0.091
Quest. 7 Successful work should be the main criterion for the advancement of every individual in society.	0.075	0.747	−0.127	0.089
Quest. 4. It is more important to me whether my civil rights are respected than what nation I belong to.	0.261	0.603	0.022	−0.197
Quest. 3 I do not value people according to which nation and people they belong to.	0.247	0.543	0.117	−0.119
Quest. 8 It is much more important to me to be accepted by my professional colleagues than by people who are particularly attached to a nation, tribe or fraternity.	−0.132	0.362	0.274	−0.321
Quest. 9 I think that the most important thing is to realize my personal interests, regardless of whether the means used for that are morally acceptable or not.	−0.117	−0.134	0.791	0.112
Quest. 10 One should listen to people who are powerful and influential in society, even if they are wrong.	−0.093	−0.047	0.775	0.223
Quest. 11 It is important for me to be accepted by friends and relatives, no matter what sacrifices my family and I have to make.	−0.121	0.035	0.759	−0.120
Quest. 17 I think that authority is authority and that we should all listen to it.	−0.029	−0.020	0.380	0.311
Quest. 1 It is important for me to be a member of the Montenegrin nation.	0.051	0.071	0.263	0.732
Quest. 2 I am very attached to my brotherhood and tribe.	0.059	−0.009	0.174	0.727

Source: Đukanović (2018), unpublished work paper.

The educated are also significantly more represented on the second factor, which is also expected, because young people are at the same time the most educated part of the population. Then it is understandable that for the same reasons why no statistically significant differences in terms of authoritarianism and nationalism were found among different age categories or among the different educational categories of the Montenegrin population.

These conclusions are indirectly confirmed by the results of double cluster analysis.

The first cluster, made up of 12.2% of the respondents, has increased values on the scale of nationalism and authoritarianism.

The second cluster, which includes 49.1% of the respondents, consists of averages that do not stand out on any factor.

In the third cluster, 23.5% of the respondents stood out. They are characterized by less interest in both the family and the profession. This points to the conclusion that this cluster consists of withdrawn and partially passive people.

Finally, the fourth cluster, which accounts for 15.2% of the respondents, is characterized by increased individuality and low ethnocentric tendencies—that is, individualists.

The first and fourth clusters were twice as important for clustering as the second and third clusters.

The results of the cluster analysis show that older men are more prone to authoritarianism and nationalism, whereas younger men, on average represented on all factors, are prone to anomie and passivity, and to a lesser extent, they also show individualistic traits. As expected, the villagers are most represented in the cluster of authoritarianism and nationalism, and the citizens of Podgorica are characterized by reduced interest in the family, business success and reduced individualism.

Unlike all aspects of everyday life researched so far, value orientations do not indicate a clear distinction between social strata, but rather, it could be said that it is atypical and then fluid and confusing. The most significant result is that members of the upper class and half of the experts are relatively more bearers of nationalist and then authoritarian tendencies, which should be expected to be more value orientations of the lower strata (only members of the subclass significantly share such orientations with the upper strata), as in the earlier socialist period, they were the bearers of dominant collectivist patterns, whether they were personal or social value orientations. In this respect, Bourdieu's theoretical views (Bourdieu, 2013) are not in line with these findings if we do not include a broader historical context. First of all, Montenegrin society (and most societies in the Balkans—the authors note) was characterized by a blocked transformation, which in terms of values resulted in normative-value dissonance (for more on this, see Lazić et al., 2018). Contrary to expectations, members of the upper classes accepted nationalist and authoritarian patterns as cost-effective from the point of view of preserving and increasing material, cultural and social capital. Truth be told, among the most educated, there is a significant percentage of people who are the bearers of individualistic-liberal tendencies, which would be in favor of Bourdieu's theory, but they are still marginal among the higher strata. That there was not a word about the true value commitments but of political conformism is best illustrated by the issue of the three most important claims on the scale of authoritarianism-nationalism and individualism, where between members of different social classes, there were no significant differences in priorities in all three selections that were on the line of individualistic value orientation.

- *Time Budget*

As a rule, the budget of time is included in the research of everyday life. In different theoretical approaches, different meanings are given to the time budget. Regardless of the fact that certain aspects of everyday life, inseparable from each other, tend to be artificially separated into separate units, we are still convinced that the time budget gives us a chronological picture of daily events that we then connect and interpret in accordance with the prevailing theoretical and methodological approach. The chronological sequence of events provided by this positivist approach, as a rule, cannot be the basis for us to understand or explain the phenomenology of everyday life, but it can be at least an auxiliary methodological tool.

The respondents are very burdened with professional and family responsibilities; family obligations take them seven hours; time spent at work with preparation and going to and from work, 9.5 hours; and they spend 2.5 hours on part-time work. In 24 hours, they have only six hours left to sleep

Observed by gender, women spend significantly less time traveling to and from work, but significantly more on household chores. Men spend significantly more time on professional responsibilities.

Older people spend significantly more free time at work and doing household chores, but also other family activities than younger ones, but because of that, younger people spend significantly more time studying, which is expected.

The more educated spend significantly less time performing family activities and work significantly less part-time than the less educated.

The lower strata spend significantly more time learning and more time doing family activities.

Basically, the division of labor of household members during the day, measured by the time spent, supports some traditional patterns, although women are in a significant percentage more engaged. As a result, they have double working hours because they did not significantly reduce family obligations by getting a job outside the home (Bourdieu, 2001). However, the specificity of Montenegro is that they gained more power in family decision-making, and with it, more symbolic power.

Another specificity is that members of the lower strata seek to increase their social and cultural capital.

Certainly, the most significant change in the traditional patterns of everyday life is the high commitment to professional obligations of all social strata.

Finally, it is interesting to note that respondents with a higher spending index spend less time on schooling and gaining higher qualifications, but also on family activities, which supports Bourdieu's thesis on layer differentiation as a key factor in the organization of everyday life and lifestyles.

- *Satisfaction With Everyday Life*

At the end of this study, we measured the degree of satisfaction of respondents in five different areas of everyday life on a 10-point scale.

High positive correlations were found between satisfaction in five different areas of everyday life as well as overall satisfaction with everyday life and stratum affiliation. This satisfaction is greatest with the job people do, their professional status and financial situation, followed by the overall satisfaction in all the above areas and finally their health and love life. In all these areas, members of the upper strata are significantly more satisfied than the lower strata.

The situation is similar between the consumption index and stratum affiliation, with the correlation with love life being the weakest ($r = 0.070$; $p = 0.037$) and health status ($r = 0.075$; $p = 0.033$) almost on the verge of significance. It is obviously about an elderly population.

As expected, young people are more satisfied in two areas than older people: love life and health status, as well as everyday life in general, whereas in all other areas of everyday life, no statistically significant differences were found by age. The more educated are also significantly more satisfied in all the above areas than the less educated, which is probably a result of stratification.

The average score of life satisfaction in these five areas is 6.166, which is slightly above the average on a scale of 1–10; the estimate of total satisfaction in these five areas is 6.8032.

We were especially interested in the relationship between the accuracy of self-perception of a certain social class and satisfaction with everyday life. It is necessary to point out that the respondents in most cases have an accurate self-perception of stratum affiliation. It is not unexpected, then, that the positive correlations between the self-perception of stratum affiliation and satisfaction with everyday life are very pronounced, even stronger than when it comes to the relationship between stratum affiliation and satisfaction. An accurate cognitive picture of belonging to a certain social class strengthens the self-awareness and self-esteem of the respondents, which enhances the feeling of personal satisfaction.

- **Strata Lifestyle**

- *Experts*

The expert stratum has the most authentic lifestyle of all the strata. At first glance, they look like an upper class, but they are fundamentally different. They are similar to the higher stratum only in terms of high consumption index. However, the high index in the stratum of experts has a completely different social symbolic meaning; whereas for members of the upper class, it symbolizes social power, prestige and reputation, for experts it is primarily about satisfying the numerous aesthetic needs generated by their strong individuality. For these reasons, more than any other layer, they show strong interest in traveling and getting to know European and non-European societies and civilizations.

The main characteristic of this social stratum is pronounced individualism. It is reflected not only in the sphere of consumption but also in family life. In the distribution of the family budget, it is known exactly what someone pays in the family, and everyone keeps the remaining part of the money for themselves. Their individualism is also reflected in the upbringing and education of younger children, which experts leave to their older daughters. Among other things, they take care of the school and extracurricular activities of the younger siblings.

Due to the emphasized individualism, they try to distance themselves from close relatives and parents because the relatives openly or covertly oppose their individualism. For these reasons, experts often come into open conflicts with close relatives, more often and more intensely than members of any other social stratum.

The pronounced individualism of experts is caused by unfulfilled professional achievements and efforts to acquire new knowledge and skills or expand old ones, in order to achieve greater competitiveness in their professions. That is why they plan professional training in the long run, yet not only because of their personal professional ambitions but also because of the improvement of their current financial position, which they are otherwise dissatisfied with. That is why they want to achieve the knowledge and skills necessary to open a private company. As they lack the financial means to do so, they seek the consent of other family members because, despite the expressed individualism, the risk seems too great. When it comes to social engagement, they are active only in those social organizations and associations with the help of which they can increase the chances for their professional promotion or which directly enable them to do so.

They are like the members of the upper stratum in that the type and content of the activities of experts in their free time is shrouded in mystery. However, as they spend most of their free time watching television and listening to the radio, one gets the impression that spending free time is in the shadow of their professional aspirations.

When it comes to cultural practices, members of this stratum, in addition to the elite culture, in which they are at least seemingly similar to the upper stratum, also prefer pop culture, which is due to the relatively larger number of young people in this than in most other strata, but they are similar to the upper class in terms of interest in political-informative programs. Due to the increased participation of young people in this stratum they reject culture to a much greater extent.

From the higher stratum, to which they are close in terms of the consumption index, they essentially differ in terms of value preferences. It is known that members of the upper class predominantly express authoritarian-nationalist value preferences, and experts in value preferences emphasize professional success and the realization of individual rights and freedoms, but also nationalism, yet to a lesser extent. This minor discrepancy in the value preferences of professionals is much more likely to be conditioned by the desire to conform to the expectations of the ruling nomenclature in order to more easily realize their professional aspirations than to the real inconsistency of their value preferences.

The lifestyle of this stratum is primarily aimed at achieving professional success and the realization of individual rights, and the other areas of their life are in the function of achieving these basic life goals. If they are not in that function, then they play a secondary role in the lifestyle of experts.

- *Small Private Owners and the Self-Employed*

Although small private individuals and self-employed have certain similarities with experts, their lifestyle still differs. Their consumption index is lower only than that of the members of the upper class, to whom they are similar in the choice of better-quality goods and the way of consumption, and in the fact that for them, consumption is primarily a status symbol. As we have seen, both layers are fundamentally different from experts, who use consumption to satisfy numerous aesthetic and cultural needs.

They also differ from experts in making family decisions; whereas in the stratum of experts, there is a consensus on independent decision-making and distribution of the family budget, small private individuals and the self-employed make family and professional decisions collectively, which brings the members of this stratum closer to the democratic type of decision-making. However, women play a dominant role in everyday decision-making. Yet, in all three upper strata of Montenegrin society, there is a tendency of increased participation of women not only in performing those family tasks that were performed by men (raising children) but also in family decision-making, unless it is about the most important family decisions that are rarely made. Elements of matriarchy are also present in the higher layers, although they are more dominant in the lower layers.

They are distinguished from experts by a much more pronounced entrepreneurial spirit, manifested in the acceptance of ever new business challenges and arrangements, but at the same time accepting numerous business risks.

Like experts, they are fans of pop culture and political-informative programs. They are also less interested in folk art. Contrary to the upper class, they are much less conformist in the manifestation of cultural practices. However, after the members of the upper class, they show the greatest satisfaction in various areas of everyday life.

It is noticeable that the emphasized entrepreneurial spirit as the main feature of the lifestyle of this social stratum is not in great harmony with important professional and family decisions, where small private individuals and the self-employed want to share business risk with close people from the family environment. This is somewhat understandable, given that material resources to correct misguided business ventures are extremely limited. Their social identity and lifestyle are not as consistent as those of the upper class and experts, and they are more reminiscent of the upper class, but their entrepreneurial spirit makes their lifestyle unique, despite everything, characterized by the greatest willingness to undertake business risk.

- *Skilled and Highly Skilled Workers*

Due to the lower consumption index, they show more deprivations in meeting different needs than members of higher social strata. They buy goods more often in Chinese stores and spend their holidays in cheaper arrangements in Montenegro.

Wives decide on the allocation of the family budget for each day. In this social layer, sociopathological phenomena, especially dependence on alcohol and tobacco, are more intergenerationally pronounced.

Skilled and highly skilled workers in the next five years do not believe that the key to advancement in Montenegrin society is hard work; they are not ready to engage in various associations and organizations, among other things due to work overload and lack of time. They show a certain ambivalence in the choice of cultural practices. They are somewhat more interested in folk art and to a lesser extent in pop and elite culture. Their value preferences are confusing and anomic; they prefer professional values and human rights to a lesser extent, as well as nationalism, but less than members of the upper classes.

The lifestyle of skilled and highly skilled workers is characterized by undeveloped tastes due to selective consumption, aimed at meeting basic needs, caution in efforts to adapt to existing unfavorable economic and professional conditions, poor future planning due to a lack of material resources and social ties, undeveloped aesthetic criteria on cultural practices, with a certain preference for folk art, but also a kind of value confusion. The feeling that they cannot change anything in an unfavorable environment is accompanied by an effort to adapt to it through hard work.

- *Unskilled and Semiskilled Workers*

The consumption index of unskilled and semiskilled workers (USW and SSW), together with a mixture of other strata, is the lowest. That is why they primarily buy the most necessary foodstuffs and personal hygiene products. When choosing, the key criterion is their price, and the place and quality of the purchase are not important to them. However, due to the low price, they are most often supplied in Chinese stores.

Women's 'patriarchy' present in the social strata analyzed so far, is still the most pronounced in this social stratum. Money from all sources is collected in a common fund, and the usual daily expenses are decided by men. When it comes to larger purchases or, possibly, investments, they rarely or never decide. It is interesting that all important family affairs are carried out or decided by the mothers of USW and SSW, from raising and educating grandchildren to determining which specific knowledge and skills the grandchildren will adopt, as well as the possibility of going on vacation. With farmers and a mixture of other social strata, they show the highest index of deprivation. USW and SSW workers are often addicted to alcohol.

Due to the lack of special knowledge and skills, they find it difficult to find the desired jobs or jobs in general, because of which they are very frustrated, and especially due to irregular income.

They are inactive in associations and organizations because they feel that this cannot be of help to them and that they can only rely on themselves.

Acquaintances of other people are primarily important to them because of the possibility of borrowing money and creating such social capital, as cultural people do not have it, nor can they create it.

In their free time, they are mostly passive. When it comes to cultural practices, they are remarkably like the SW and HSW. Their value orientations are not clearly profiled, but they are less prone to authoritarian and nationalist value orientations than members of the upper classes.

Overall, their lifestyle is characterized by constant confrontation with financial and professional problems, against which they feel powerless, passive and anomic.

Cultural spending is relatively weak, vague and confusing, and in leisure time, they show marked passivity. Their value orientations are undifferentiated and confusing, although authoritarian-nationalist patterns are not close to them. Consequently, they are passive and distrustful of social organizations, which results in their pronounced rejection, followed by feelings of hopelessness and anomie to these organizations being strengthened and thus relying only on themselves.

- *Mixture of Other Strata*

This 'stratum' is a kind of mixture of different social groups (pensioners, housewives, pupils, students, etc.), which is most connected by the fact that they are dependent social categories. In many other ways, they are different, often contradictory, which certainly conditions the inconsistency, the incoherence of the lifestyle of this social 'stratum'.

Different social groups of this 'stratum' have in common an extremely low consumption index. In addition to the extremely low consumption index, the USW and SSW and farmers have in common a high deprivation index. Like the mentioned strata, members of this 'strata' buy only the most necessary products for nutrition and personal hygiene, at the lowest possible prices.

As their budget is small and uncertain, they do not plan distribution in the long run, but in accordance with current needs. Grandmothers play a dominant role in making decisions about distribution, and grandfathers much less often, and the decision about a possible vacation as well as a place of rest is also made by grandmothers. Due to the dominance of women in this 'stratum', other family members come into conflict with women. These strong conflicts have an intergenerational basis because members of this 'stratum' are significantly more likely than other strata to come from families of divorced parents. Only in the professional sphere do men decide independently, which is probably conditioned by the presence of a larger number of young people (e.g., students) who believe that the choice of profession is their sovereign right.

They do not try to improve their bad economic situation with self-provision, even the young people. They accept it passively, almost fatally. If they possibly have some extra work, they do not want to do other jobs. They are ready to leave a job only if the salary is high.

Children of this 'stratum' do not possess special knowledge and skills, and they cultivate social ties only if they are in the function of improving their health. We should not lose sight of the fact that pensioners are the most represented in this layer.

Unlike the lower social strata, the members of this 'layer' are nonconformists, which is probably a consequence of the increased presence of young people.

They are predominantly interested in pop and shun culture (again due to the greater presence of young people) and to a lesser extent in political-informative programs and elite culture (older people). When it comes to value preferences, they prefer family values, professional success and individual rights (profession and rights, probably due to the presence of young people). Focusing on family values is a kind of compensation for traumatic experiences from parental families. At the same time, members of this class (pensioners and housewives) show certain authoritarian value preferences that are weaker in intensity than that of the higher stratum. We see that in most areas the daily life of the members of this 'stratum' is inconsistent and even contradictory, which is due to the great sociodemographic diversity, due to which the members of different generations are often in latent or open conflict.

They point out corruption as the main cause of their great deprivation and see the only way out in mass riots, which is again connected with a larger presence of young people. Discontent due to social frustrations is 'resolved' in very pronounced intra-family conflicts, which are intergenerationally transmitted in this 'stratum'.

They are passive at leisure, but in value orientations, they are more oriented toward family and work than members of other social strata. However, they also show some authoritarian-nationalist tendencies. Their cultural spending is diverse, although it tends somewhat more toward elite culture (student presence). In the system of cultural practices, value preferences, their lifestyle, similar to officials and technicians, is characterized by contradictions and confusion and, in relation to society, by extreme rebellion.

9.1.5 Conclusion

The higher and lower strata differ most clearly and the most in the consumption index. The first element of these differences are allocations for food, housing and utilities. Our analysis showed that these expenditures are highly positively related to the position of the stratum on the social scale and the height of the consumption index.

Members of the higher strata in the choice of goods favor the quality, variety and aesthetic appearance of the product irrespective of the price. They mostly buy goods in large shopping centers and in specialized shops with branded goods. If there are no such centers and shops in their city, they buy goods in other cities of Montenegro, Podgorica, in neighboring countries or throughout Europe. Quite often, they order food from better restaurants. They spend their holidays in foreign luxury hotels. Members of the higher strata, by choosing and places of purchase, try to demonstrate the built tastes that 'oblige' them according to the stratum affiliation.

In the lower strata, the situation is almost reversed; the variety, quality and aesthetic appearance of the goods are secondary to the price. That is why they buy goods in Chinese shops, flea markets or in the nearest stores, especially when it comes to groceries. In general, consumption is in the function of satisfying basic biological needs, which is conditioned by a significantly lower consumption index than the upper strata.

It should not be lost sight of the fact that members of the upper strata are significantly more often members of the ruling nomenclature or are connected with it by numerous social ties, whereas members of the lower strata significantly more often do not belong to any social or political group.

Overall, the lifestyles of the social strata are primarily determined by stratum affiliation. Diversity, content and richness of lifestyle are in a high positive correlation with the position of the stratum in the stratification of Montenegrin society. If we say that the higher strata have a unique lifestyle in a purely didactic sense, then we could say that it is more diverse, richer and more meaningful in all the examined areas than the lifestyle of the lower strata. These synthetic indicators fully support Bourdieu's Tory concept. The baseline of separation is layered. As we have marked the level of education as a key indicator of cultural capital, education as well as stratum affiliation are key elements of the distinction between strata, simply because there is a high positive correlation between education and stratum affiliation.

In general, cultural practices are differentiated according to the stratum and educational level, but cluster analysis showed that cultural practices among the strata also more or less contain a mixture of conflicting contents, which is especially the case with value preferences. Thus, the higher stratum, even the stratum of experts, show authoritarian-nationalist value orientations, although that would be least expected of them. Members of the lower classes are more inclined to democratic value orientations, which is also not expected, although their tendency toward egalitarian value preferences is.

The strengthening of egalitarian tendencies among spouses can be observed in the entire sample of respondents, because women in all strata to a lesser or greater extent gain a greater role in making various family decisions. Only at first glance, no significant differences were observed between the layers, and it could be said that Bourdieu's theory in this case does not provide the possibility of explication. However, it can be noticed that whereas in the higher strata, despite certain deviations, there is greater closeness and joint decision-making, especially when it comes to making the most important family decisions, in the lower social strata, some anachronistic patterns of 'matriarchy' are more pronounced and have a decisive voice when there is disagreement in their adoption. If we consider this broader cultural-historical context, Bourdieu's theory seems acceptable to us in this case as well.

When it comes to atypical value preferences of social strata, which have been talked about a lot before, is there any basis to consider Bourdieu's theoretical concept inapplicable? Again, considering specific sociohistorical factors, there are not enough grounds to reject Bourdieu's theoretical views in this case either. These factors when it comes to value preferences are best explained by Lazić's theory of value-normative dissonance (Lazić et al., 2018). Professor Lazić believes that Montenegrin society has necessarily experienced cognitive dissonance due to the blocked transformation and adds: 'The problems of harmonizing the new social order and dominant value orientations in Montenegro are all the more serious because the same tendency to strengthen value-normative dissonance and value confusion (spreading skepticism towards liberal values, characteristic of the capitalist order; maintaining values typical of the socialist order; simultaneously supporting the elements of both value systems, among members of all fundamental social groups) occurs in both fields, political and economic' (Lazić et al., 2018).

The verification of Bourdieu's theory is possible only in the examples of developed and stable societies, because only in them certain categories of fields (capital, habitus, etc.) can be developed and recognized. This is not the case with Montenegrin society.

9.2 Social Prospects of Montenegro: Indexes

In order to gain a detailed insight into the current situation in Montenegro, in addition to the original research paper, we have at our disposal ready and available analyses of various institutions such as various working groups, analytical centers or 'think tanks', private profit and nonprofit organizations and reports of official international bodies. In the next segment, we will pay special attention to this type of material and show several different indicators and indices; we will determine to which implications they lead individually and what is the general image if they are observed together—that is, what is the profile of the country they make.

9.2.1 Democracy Index

The first indicator we will consider is the democracy index, which is calculated each year by the Economist Intelligence Unit (EIU), the development and analysis department at the Economist Group. The similarity of the name of this group with the famous magazine the *Economist* is not accidental as the *Economist* newspaper and the Economist Group are sister companies. When creating the index of democracy, one must keep in mind that the very notion of democracy is very vague and that there is no one generally accepted definition of this term. Therefore, every attempt to measure and number is a problem, because how to assign a numerical value to something that is not clearly defined (Coppedge, 2012)? Therefore, the authors of various measures of democracy must first define their starting position and set their definition of democracy, or clearly refer to an already existing definition, which they can further use as a starting point. In its approach, the EIU does not actually give an accurate definition of democracy but starts from and builds on preexisting measures of democracy. Thus, for example, the measure of the degree of democracy made by 'The Freedom House' is taken as well as their definition of democracy, which relies on the degree of presence of political freedoms and civil rights. The authors believe that these characteristics are the basis of any definition of democracy, but that they alone are not sufficient to understand the degree to which democracy is present and widespread in a society (EIU, 2019). Therefore, the authors have expanded such a concept; currently the democracy index consists of five factors, and they are: electoral process and pluralism, functioning of state administration, participation in political life, political culture and civil liberties. According to the results on these factors, states are further classified into full democracies, flawed democracies, hybrid regimes and authoritarian regimes (EIU, 2019). The classification into categories is made according to the total number of points and is therefore not sensitive to variations in scores on individual factors. Therefore, it is mathematically possible for a country to have an extremely low score on some of the factors and an extremely high score on all the others and to be assessed as highly democratic. Of course, this will not be seen in practice because there is a high degree of correlation between factors, and the chances of such discrepancies occurring are practically nonexistent. When interpreting the results of the democracy index, it should be borne in mind that the degree of democracy is not only a political and cultural indicator but there are strong implications that democracy is associated with economic success and welfare of the state and that its decline has negative effects on the economy

(Acemoglu, Naidu, Restrepo, and Robinson, 2018). Tables show the indices for Montenegro as well as for the surrounding countries. Croatia and Slovenia are singled out as they became members of the European Union but are relevant as former republics within Yugoslavia.

Table 9.3 shows that Montenegro has the second-lowest score (higher scores indicate a higher degree of democracy). As expected, Slovenia has the highest score of 7.5, followed by similarly ranked Croatia and Serbia. These three states are also the only one that have the status of democratic states—flawed democracies. Albania and Montenegro follow, followed by Bosnia with a score of just 4.86. Therefore, according to the assessment of this index, Montenegro is currently one of the least democratic countries in the region and belongs to the category of hybrid regimes. What does that mean? Hybrid regimes are defined as those where numerous irregularities in the electoral process prevent free and fair elections and where pressure on opposition parties and politicians is common. This situation is usually accompanied by widespread corruption, weakened rule of law, government influence on the judiciary and widespread pressure on journalists and the media. To gain a better insight into this result, these countries were compared by individual factors.

In Table 9.4, it is especially interesting to compare Serbia and Montenegro, as they coincide on two indicators: namely, the functioning of the state

Table 9.3 Democracy index for Western Balkan countries and country type

	Index—2019	Country type
Montenegro	5.65	Hybrid regime
Serbia	6.41	Flawed democracy
B&H	4.86	Hybrid regime
Albania	5.85	Hybrid regime
N. Macedonia	5.97	Flawed democracy
Croatia	6.57	Flawed democracy
Slovenia	7.5	Flawed democracy

Source: Author based on: EIU (2019).

Table 9.4 Factors of democracy index for selected countries

	Montenegro	Serbia	B&H	Albania	N. Macedonia
Electoral process and pluralism	5.67	8.25	6.17	7	7
Functioning of government	5.36	5.36	2.93	5.36	5.36
Political participation	6.11	6.11	5.56	4.44	6.67
Political culture	4.38	5	3.75	5	3.75
Civil liberties	6.76	7.35	5.88	7.65	7.06

Source: Author based on: EIU (2019).

administration and participation in political life. Then where is Montenegro lagging? The first and most obvious difference is the factor of electoral process and pluralism, where Montenegro has by far the lowest score compared to all countries in the region. This factor covers issues such as the ability to vote freely without fear, the equality of political parties in the electoral process and the like, and the low score is worrying. It can also be seen that Montenegro lags in political culture, but the difference is smaller, and it is not ranked the worst. The situation is similar on the indicator concerning civil liberties.

To better understand the current situation, it is necessary to have an insight into how it came about. Fortunately, this index has been around for many years, and the results for previous years are publicly available, so creating a trend view is possible. The chart clearly shows that the degree of democracy of Montenegrin society has been continuously declining for more than a decade.

Of course, it is necessary to point out in the end that this index is created as an assessment of a group of experts, but that the process of making scores is not completely transparent, and it is impossible to assess the exact level of objectivity. On the other hand, Figure 9.1, democracy index of Montenegro for the period 2006–2019, shows the fact that the democracy index of Montenegro is declining despite the continuous orientation toward Western Europe and America (which culminated in joining NATO) leading us to the conclusion that the creators of the index are not too influenced by pro-Western favoritism (except for the fact that the definitions of democracy

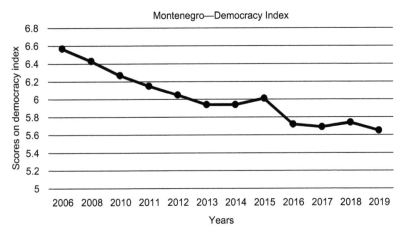

Figure 9.1 Democracy index of Montenegro for period 2006–2019

Source: Author, based on: EIU (2019).

9.2.2 Corruption Perception Index

The next index deals with the level of corruption in the country (corruption perception index). This index is compiled by the nonprofit organization Transparency International, from Berlin. This organization defines corruption as 'abuse of public office and power for personal gain' (Corruption Perception Index, 2019). Some definitions of corruption are limited to the illegal use of power for personal gain. Such definitions are too narrow, and as Daniel Kaufman observes, such a definition does not take into account that someone in power can legitimize the actions he or she takes and thus exclude himself or herself from the previous definition of corruption (Hellman, Jones, and Kaufmann, 2003). Thus, the definition used by Transparency International includes both types of corruption, illegal and legal. It can be noticed in the name of the index that we are talking about the perception of corruption and not about corruption itself. This choice of name for index is based on the fact that corruption itself is very difficult to measure by concrete measures and must rely on the assessments, feelings and visibility of corruption. In order to get the most accurate picture, the results of assessments of a large number of agencies and associations such as the World Bank and the World Economic Forum are collected as well as the results of various surveys and research. For a country to be included in the index, it is necessary that there be at least three different sources of information for it. The corruption perception index as a measure proved to be valid through a study that compared this measure and other generally accepted indicators of the presence of corruption, such as the presence of black and gray markets and the burden of the state apparatus with too many regulations (Wilhelm, 2002). As in the previous case, the results for Montenegro are presented, followed by the results of the countries in the region. When interpreting the results, it should be borne in mind that the results can range from 0 to 100, with higher results indicating a lower level of corruption.

Table 9.5 shows that Montenegro has one of the better results in the region and that according to this indicator it stands side by side with Croatia ahead of all other countries with the exception of Slovenia. With a score below 50, Montenegro is entering the group of countries considered to be countries with corruption problems. According to this indicator, Montenegro is therefore one of the best placed in the region, but is there a growing trend of corruption or is the perception of it declining?

Table 9.5 Corruption perception index for Western Balkan countries

Countries	Index—2019
Montenegro	45
Serbia	39
B&H	36
Albania	35
N. Macedonia	35
Croatia	47
Slovenia	60

Source: Author, based on corruption perception index, 2019.

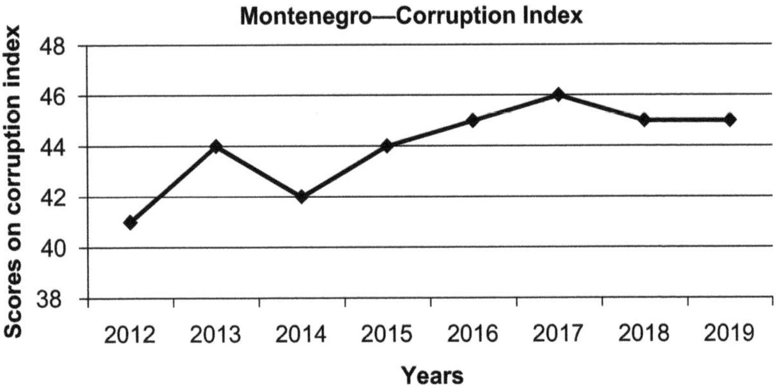

Figure 9.2 Corruption perception index of Montenegro for period 2012–2019
Source: Author, based on corruption perception index, 2019.

As can be seen from the attached chart, there is a general trend of reducing corruption in Montenegro, which reached its peak in 2017, when there was a leveling and stagnation of results. It can therefore be concluded that there was a trend of improvement, but which has currently stopped, and that work needs to be done to reestablish the trend. We would like to mention once again that these indices are indicators of the perceived state and not empirically determined indicators, and as such, they can be subject to prejudices and similar errors in the process of formation.

9.2.3 Global Health Security Index

To assess the robustness of health systems and their resilience to potential epidemics and pandemics, in 2019 the global health security index was developed. Behind this index are the Johns Hopkins Centre for Health Security, the Nuclear Threat Initiative and the aforementioned Economist Intelligence Unit. In addition to the members of the aforementioned organizations, 23 experts from relevant fields from 13 different countries participated in the creation of this index. The index itself is based on 140 different questions that are classified into 85 subcategories, 34 indicators and six general categories (Global Health Security Index, 2019). The six categories are as follows: prevention: prevention of the emergence and spread of new pathogens; detection and reporting: early detection and reporting of epidemics that may affect the international community; rapid response: rapid response to the epidemic and its suppression; health system: quality and robustness of the health system and its ability to protect health workers; adherence to international norms: willingness to work on improving national capabilities to deal with such situations, practical and financially covered plans to cover existing shortcomings and adherence to international standards; environmental risk: general level of risk and vulnerability of the state to biological threats. This index was published in October 2019, just before the outbreak of the epidemic and later the COVID-19 pandemic, so the assessment of experts was immediately tested in practice. In addition to the review of regional scores and a more detailed insight into the result of Montenegro, we will also examine how the predictions of experts turned out.

The global health security index and its factors for the Western Balkan countries (2019) indicates that Montenegro, among the reference countries, ranks right in the middle. Albania and Serbia occupy first-place positions, whereas Bosnia and Herzegovina and Northern Macedonia rank worse. The lowest grade and at the same time the most worrying is the grade of the health system. The wide accessibility of the public health system is considered a plus, but other factors that make up this score are poorly evaluated, with the poor score on the assessment of the capacity of hospitals and other health care institutions being especially noteworthy (the score of 16.9, which is lower than the world average of 24.4). The second-lowest score realized by Montenegro is on the quick response factor, where it achieves the lowest score in the region. According to the estimates of the experts who made this index, Montenegro completely lacks a plan for the fight against a serious epidemic, and therefore, there are no exercises to prepare for such a situation. On the other hand, the access of the population to communication channels through which they might be warned of a possible danger and through which the health system can issue instructions to citizens scores well. On the positive side, in assessing the riskiness of the environment, Montenegro is among the best placed in the region, which is due to good (but still not excellent) assessments of political and socioeconomic stability. The second-best rated factor is the ability for quick detection. The individual score that significantly contributes to such a good assessment is the assessment that Montenegro belongs to a group of countries that have elaborate mechanisms for connecting various relevant services and consolidating their data. It is interesting to note that Montenegro has a higher score on adherence to international norms than all reference countries, and that score is higher than the score achieved by Croatia, although it has largely become a member of the European Union. To examine the validity of the index, results were taken for 10 regional countries: Montenegro, Serbia, Northern Macedonia, Bosnia and Herzegovina, Croatia, Slovenia, Greece, Romania and Bulgaria. These countries were chosen because they are regionally close, and therefore, the virus appeared in them within a relatively short span. In order to be able to compare the data, the data related to a certain day were not taken, but the number of infected and dead on the 45th day after the first detected case was taken for each country. The results were taken from the site of the Johns Hopkins Centre, which was chosen also because he participated in the creation of the index in question. Using these numbers and population numbers (sourced from the World Bank database), the infection and mortality rate per 1,000 citizens was created, and mortality rate for the infected was calculated. These measures were further correlated with the overall index and each of the factors. The correlations obtained were high and generally ranged from 0.2 to 0.5, but probably because of the small sample, they were not statistically significant. It is important to note that these correlations, contrary to expectations, were not negative but positive—that is, higher scores were associated with a higher degree of infection but also of mortality. Mortality per 1,000 inhabitants had a significant correlation only with quick response, $r = 0.676$; $p = 0.032$. So the general impression is that the higher indices indicate a potentially worse reaction. Of course, this result should be interpreted with caution due to the small sample, and it is necessary to do a more detailed study involving a larger number of variables and countries.

9.2.4 Social Progress Index

How to determine exactly how prosperous a country is? The approach most often taken as a benchmark is GDP, but several experts consider this method to be incomplete because it considers only one type of economic indicator, yet ignoring other relevant quality-of-life indicators. In response to this problem, a number of different indicators and indices have emerged, one of which is the social progress index (Sen, North, and Stiglitz, 2013) This index is created by the nonprofit organization Social Progress Imperative. According to the organization itself, the index was created as an attempt to quantify aspects of the degree of development in order to identify areas that need to be improved and to which attention should be paid in each individual country (Social Progress Index, 2019). The index covers three broad areas: basic human needs, foundations of well-being, availability of Opportunity. Each of the areas is further divided into subareas that are then divided into individual indicators. These indicators are partly objective statistics such as 'the chance that a child will die by the age of five' (basic human needs, subindicator—nutrition and basic medical care) and are partly estimates by experts from the country to which the index refers. The transparency of this project is also commendable, as all data are available free of charge, and in the data file, for each individual indicator, a reference is given that is used in its implementation.

Table 9.6 shows the data for 2019 where the total index is shown as well as each of the three factors. The absence of results for Bosnia and Herzegovina can be immediately noticed in the table. Not all data through which the indices are derived are available for this country, so it does not have a result in the overall index and is therefore excluded from the analysis. It can also be observed that the results for Montenegro and the three reference countries are similar, especially in the general score, which is actually the average score of the three individual factors. The most critical point, both for Montenegro and for the entire region, is the opportunity factor.

A more detailed examination of individual indicators of the chance and opportunity factor shows that the problems in Montenegro are multiple. First, there are low scores on political rights (23 points out of 40) and limited freedom of expression (0.64 where 0 is a complete lack of freedom and 1 is complete freedom). Further, there are factors such as nonacceptance of sexual minorities and discrimination and violence against minorities (8.5 out of 10!). The last factor that significantly reduces the score of Montenegro and the region is the number of universities that are ranked globally as well as the percentage of students within the country who attend them (0 is the score on both indicators).

Table 9.6 Social progress index and its two main factors for Western Balkan countries, 2019

	Social progress index	Basic human needs	Foundations of well-being	Opportunity
Montenegro	71.16	84.95	76.88	51.66
Serbia	71.59	86.00	70.97	57.81
Albania	71.57	85.03	79.03	50.65
N. Macedonia	68.92	86.37	72.81	47.57
Croatia	79.21	90.90	80.88	65.86
Slovenia	85.80	95.64	86.18	75.58

Source: Author, based on social progress index, 2019.

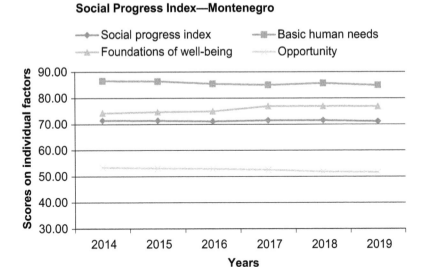

Figure 9.3 Social progress index and its main factors for Montenegro for 2014–2019
Source: Author, based on social progress index, 2019.

This chart shows how the results for Montenegro have been developing since 2014, when the index was first published, until 2019. As can be seen the general impression that the results give is a general stagnation where basic human needs have a slight downward trend and foundations of well-being an equally slight upward trend, so the average index remains unchanged.

9.2.5 Global Multidimensional Poverty Index

In 2015, the United Nations General Assembly adopted 17 goals for sustainable development. The first of these goals is the elimination of poverty in the world. To track progress toward this goal, Oxford University, in collaboration with the Human Development Report Office, has jointly developed an index that tracks multiple aspects of poverty and deprivation in health, education and general living standards across 10 different indicators. This approach should provide a better insight into the areas most affected by poverty and complement the measure of the general poverty rate measured by the number of people living below $1.9 (Figure 9.4) per the social progress index and its main factors for Montenegro for the period 2014–2019 on a daily basis (Oxford Poverty and Human Development Initiative, 2019). It is interesting to note that a higher level of poverty does not necessarily lead to greater dissatisfaction with life; dissatisfaction occurs

Measuring Quality of Life 143

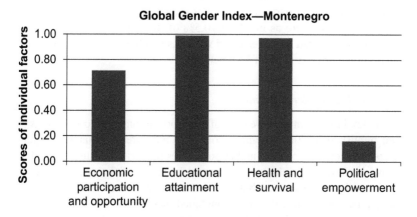

Figure 9.4 Main four factors of global gender gap index for Montenegro, 2020
Source: Author, based on World Economic Forum (2020).

when there is inequality in the community on one of the 10 factors—that is, if a part of the population is affected and a part is not. If everyone in the community is affected by some form of poverty, it does not reflect badly on the general satisfaction with life (Strotmann and Volkert, 2016). Since this index was not derived for each country in the same year but the indices were derived for different years in the period 2007–2018, it is not entirely possible, as in the example of previous indices, to compare countries in the region as their indices were not derived in the same years. For example, it is not appropriate to compare the results of one country from 2008 during the peak of the global economic crisis with the results of another country in 2018. Further, as only one review was performed for most countries, we are not able to review how the result for Montenegro has changed over time. The index of Montenegro is 0.002 (Armenia has a minimum of 0.001, and Niger has a maximum of 0.590), which is a very good value compared to all other countries on the list, and only six other countries have a smaller index. It should be noted that not all countries are present on the list and that only developing countries are included (Oxford Poverty and Human Development Initiative, 2019). Montenegro has 0.4% of the population that falls into the definition of multidimensional poverty, and 0.1% falls within the framework of highly pronounced poverty. In addition to this part of the population, it is also estimated that about 4.3% of the population is at risk of being in the group of the poor. Also, the assessment of the experts who made this list is that there are no people in Montenegro who live on less than

$1.9 a day. The extent to which Montenegro and the region are generally ahead of other poorer parts of the world is shown by the processes for developing countries where as much as 10.5% of the population in the group are severely affected by poverty and where as much as 14% live below the international poverty limit of $1.9 per day.

9.2.6 Global Gender Gap Index

The index in question is a creation of the World Economic Forum and was first published in 2006. Its main goal is to measure and quantify differences in access to resources and opportunities that arise because of gender. Thus, the index as such is not a general indicator of the accessibility of opportunities and resources, but how unevenly they are distributed between the genders. From this, it can be concluded that in case one country has a lower score than another, it does not necessarily mean that women from that other country have less access to resources but only that the access ratio is worse. Thus, we can have countries with excellent economic conditions with a lower score than more equal but less developed countries. One example would be the comparison of Montenegro and Saudi Arabia, where Saudi Arabia ranks 146th and Montenegro 71st, with Saudi Arabia having much better economic indicators. It is important to note that this index is focused on the poorer position of women and that it is not really a true indicator of equality, because almost every indicator has any score in favor of women—that is, women being in a privileged position are considered as actually achieving complete equality on that indicator. The authors of the index justify this approach by saying that they do not want one or more scores in favor of women to reduce another bad score by their influence (World Economic Forum, 2020). The general index is divided into four factors: economic participation and opportunity, educational attainment, health and survival and political empowerment. These four factors are further calculated from 14 individual indicators, 13 of which are based on publicly available and objective quantitative data.

From Table 9.7, we can see that Montenegro has the lowest value of this index in the entire region, but it should be borne in mind that the differences between all countries are small and should be measured to the second decimal place. Further, it is interesting to note that the indices of the two members of the European Union, unlike the previously examined indices, are either the same rank or lower than the non-EU countries in the region. Having in mind the individual factors that make up the Montenegrin score, we can conclude that there is a significant imbalance between them. Thus, on the factors related to education and health, Montenegro achieves an almost perfect result, but in the sphere of political empowerment of women, it shows serious shortcomings. The economic participation factor has a good

Table 9.7 Global gender gap index and its two main factors for Western Balkan countries, 2020

Countries	Index—2020
Montenegro	0.71
Serbia	0.74
Albania	0.76
N. Macedonia	0.71
Croatia	0.72
Slovenia	0.74

Source: Author, based on World Economic Forum (2020).

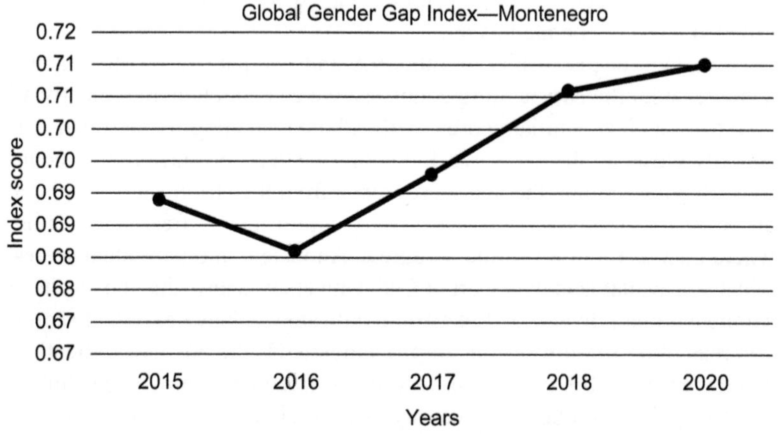

Figure 9.5 Global gender gap index of Montenegro for the period 2015–2020

Source: Author, based on World Economic Forum (2020).

score but indicates opportunities for further improvements. The low score achieved by Montenegro on women's political empowerment probably lowers the overall score achieved because research by Sharon Mastracci (Mastracci, 2017) indicates that this factor is most influential on the overall score and that promoting women at public functions is the easiest way to improve the general index score.

Observing the results of Montenegro for the period from 2015 to 2020, we can see that there is generally a slight growth trend.

9.2.7 Human Freedom Index

The human freedom index was created in collaboration with the Cato Institute, the Fraser Institute and the Friedrich Naumann Foundation for Freedom. Like other presented indices, this one was created as an attempt to quantify the indicators of one phenomenon and express them through a clearly defined index in order to draw attention to potential problems, aspects within countries that need to be improved and to enable direct comparison of countries. For the purposes of this index, freedom is defined not by an affirmative deficit that would specify what constitutes freedom, but by a negative definition—that is, freedom is considered to be the absence of restrictions and coercion that would affect an individual or institution. This definition of freedom is based on the philosophical tradition of which the most important representative is John Locke, who defines freedom as a lack of submission to the will of others and the ability to freely follow one's own will (Smith, 2013).

The general index consists of two aspects—namely, personal freedoms and economic freedoms, which further consist of 76 individual indicators grouped into 12 separate factors.

If we look at the results in Table 9.8 and compare what Montenegro achieves with the surrounding countries, we can conclude that mostly all countries achieve similar scores, with Albania standing out and which is actually closer to the separated EU countries than the surrounding countries. Bearing in mind that the scores range from 0 to 10, it can be said that all countries, except Albania, enter the group of mostly free countries but where there is room for serious progress. An insight into individual scores, which are available in the report of Cato University, shows that the results of Montenegro are diverse and that some indicators achieve excellent scores, whereas other require serious improvements. Of the positive indicators, Montenegro has the two highest scores on freedom of movement (maximum 10 points) and personal security (9.1 out of 10). The two worst scores, which are very indicative of problems in the functioning of the entire state, are the legal system and property rights (4.9 out of 10) and the rule of law (5.2 out of 10).

Observing the movement of the freedom index for Montenegro over time, we can see the downward trend, which is primarily driven by the decline of

Table 9.8 Freedom index for Western Balkan countries, 2019

Country	Index—2017
Montenegro	7.43
Serbia	7.3
B&H	7.37
Albania	7.84
N. Macedonia	7.34
Croatia	7.86
Slovenia	7.97

Source: Author.

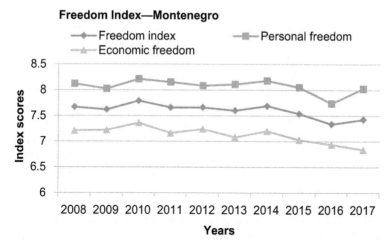

Figure 9.6 Freedom index of Montenegro for 2015–2020
Source: Author.

economic freedoms, whereas personal freedoms are mostly stable, except for the decline in 2016, which returns to regular levels next year.

9.3 Conclusion

The overview of selected indices shows the picture of Montenegro as a country belonging to the group of developing countries and that faces challenges that are characteristic of this type of country but also of the geographical region in general. Having that in mind, what would the profile of Montenegro look like? It is a state in which democracy is in crisis and whose degree of democracy is declining over time and is characterized by a lack of equality in the electoral process that the government uses to maintain positions of power. Despite this, the problem of corruption is gradually diminishing. The global health security index points to Montenegro's poor preparedness to deal with unexpected epidemiological challenges, which is in fact a reflection of the lack of proactivity and integration of various state bodies. Of course, having in mind the currently active COVID-19 pandemic, we are attending a practical test of the experts' assessment, and the initial data indicate the possibility that their assessments are unfounded. Although Montenegro does not belong to the richest countries, it is not exposed to high poverty rates, and according to world standards, it is well placed, but it must be noted that there is plenty of room for improvement, and if poverty is not a problem, it must be noted that according to quality-of-life indicators, it is felt especially lacking in chances and perspective. Also, in accordance with the previous characteristics for Montenegro, it must be noted that it is a country in which the expression of freedoms is slowly declining, primarily, economic freedoms. On the issue of gender, Montenegro is achieving good results in general, but almost completely fails in the political empowerment of women. If all the features are summed up, it must be concluded that there is one problem that runs through the whole picture and that is the underdevelopment of the political system—that is, the ruling party currently actively suppresses competition and strives for centralization, which is manifested through influences on electoral process and reduction of economic freedoms. On the other hand, at first glance, perhaps paradoxically, such a situation leads to a gradual reduction of corruption. Such a finding would not be inexplicable. The first possibility is, as the authors of the index pointed out, a government that has strong control can actually legalize certain aspects of corruption and thus soften the perception of the people and perhaps the experts on the extent of the phenomenon. Another option is for the centralized and long-standing government in present-day Montenegro (currently the ruling party has been in power in Montenegro since the introduction of the multiparty system) to control who can take how much, which potentially reduces 'independent' corruption. Of course, all of the above interpretations are hypotheses at best and should not be considered as claims, and it is necessary to approach the topic in much more detail, which is beyond the scope of this study. Finally, we must draw certain parallels between the conclusions drawn in the exploratory study and the picture presented by world experts through various indices. Having in mind the tendency of declining democraticity and increasing concentration of power in the hands of the government, the study noted the division between social classes, certain signs of danger of creating a class-divided society and, in the worst case, a pseudo-caste society are shown. If we know that classes differ and that the upper class is intertwined with the ruling structures, then its increasing centralization of power is worrying, which will certainly have negative consequences on vertical social mobility, in line with the reduced chances and opportunities that the average Montenegrin has according to the presented index. This tendency is seen in the observed greater democracy of the lower classes, opposing the authoritarian elite, which is increasingly limiting it.

Note

1. The deprivation index was calculated as follows: the number of deprivations per respondent, which was then summed up and divided by the number of respondents.

References

Acemoglu, D., Naidu, S., Restrepo, P., and Robinson, J.A. (2018). Democracy does cause growth. *Journal of Political Economy*, 127(1). https://doi.org/10.1086/700936

Allat, P. (1993). Becoming privileged. In I. Bates and G. Riseborough (Eds.), *Youth and Inequality* (pp. 139–159). Milton Keynes: Open University Press.

Bennett, T., Savage, M., Silva, E.B., Warde, A., Gayo-Cal, M., and Wright, D. (2009). *Culture, Class, Distinction*. London: Routledge.

Bernstein, B. (1971). *Class, Codes and Control* (Vol. 1). London: Routledge.

Bolčić, S. (Ed.). (1995). *Social Changes and Everyday Life*. Belgrade: ISIFF.

Bourdieu, P. (1990). *In Other Words: Essays towards a Reflexive Sociology*. Stanford, CA: Stanford University Press.

Bourdieu, P. (1997). *Language and Symbolic Power*. Cambridge: Polity Press.

Bourdieu, P. (1998). The myth of globalization and the European welfare state. In *Bourdieu Acts of Resistance* (pp. 29–44). Cambridge: Polity Press.

Bourdieu, P. (2001). *Masculine Domination*. Podgorica: CID i Univerzitet Crne Gore.

Bourdieu, P. (2005). *The Social Structures of the Economy*. Cambridge: Polity Press.

Bourdieu, P. (2013). *Distinction: A Social Critique of the Judgement of Taste*. Podgorica: CID i Univerzitet Crne Gore.

Bourdieu, P. and Passeron, J. (1977). Reproduction in education, society and culture. *Word*, 82(28): 67–121.

Cohen, P. (1998). Replacing housework in the service economy: Gender, class, and race ethnicity in service spending. *Gender & Society*, 12: 219–231.

Coleman, J. (1988). Social capital in the creation of human capital. *American Journal of Sociology* (supplement), 94: 95–120.

Coppedge, M. (2012). Defining and measuring democracy. *Democratization and Research Methods*, 11–48. https://doi.org/10.1017/cbo9781139016179.002

Cvejić, S., Radonjić, O., Kokotović, S., Vujović, S., Babović, M., Petrović, M., Mojić, D., and Spasić, I. (2010). *Living With Reforms: Citizens Facing Challenges of Transition*. Belgrade: Čigoja štampa i ISIF.

Douglas, H. (1998). Does cultural capital structure American consumption? *Journal of Consumer Research*, 25: 1–25.

Đukanović, B. (2018). Everyday life and lifestyles of social stratas in Montenegro. In D. Vukčević et al. (Eds.), *Social Cross Section of Montenegrin Society* (pp. 102–337). Podgorica: CANU.

Economist Intelligence Unit. (2019). *Democracy Index 2019*. London: EIU.

Gershuny, J. (2000). *Changing Times: Work and Leisure in Postindustrial Society*. Oxford: Oxford University Press.

Giddens, A. (1991). *Modernity and Self-Identity: Self and Society in the Late Modern Age*. Cambridge: Polity Press.

Gupta, S. (2006). Her money, her time: Women's earnings and their housework hours. *Social Science Research*, 35: 975–999. https://www.sciencedirect.com/science/article/abs/pii/S0049089X05000323

Gupta, S. (2007). Autonomy, dependence, or display? The relationship between married women's earnings and housework. *Journal of Marriage and Family*, 69: 399–417.

Gupta, S., Evertsson, M., Grunow, D., Nermo, M., and Sayer, L. (2009). *Housework, Earnings, and Nation: A Crossnational Investigation of the Relationship*

Between Women's Earnings and Their Time Spent on Housework. University of Massachusetts-Amherst. Unpublished manuscript (accessed on June 24, 2009).

Hellman, J.S., Jones, G., and Kaufmann, D. (2003). Seize the state, seize the day: State capture and influence in transition economies. *Journal of Comparative Economics*, 31(4): 751–773. https://doi.org/10.1016/j.jce.2003.09.006

Ignatow, G. (2009). Culture and embodied cognition: Moral discourses in internet support groups for overeaters. *Social Forces*, 88(2): 643–669.

Joas, H. (2000). *The Genesis of Values*. Chicago: University of Chicago Press.

Katz-Gerro, T. (2002). Highbrow cultural consumption and class distinction in Italy, Israel, West Germany, Sweden, and the United States. *Social Forces*, 81: 207–229.

Lamont, M. (1992). *Money, Morals and Manners: The Culture of the French and American Upper-middle Class*. London: University of Chicago Press.

Lazić, M. et al. (2018). Social mobility. In D. Vukčević et al. (Eds.), *Social Cross Section of Montenegrin Society* (pp. 11–54). Podgorica: CANU.

Mastracci, S. (2017). The effect of women's representation on the Global Gender Gap Index. *International Journal of Public Sector Management*, 30(3): 241–254. https://doi.org/10.1108/ijpsm-05-2016-0095

Miles, A. (2014). Addressing the problem of cultural anchoring: An identity-based model of culture in action. *Social Psychology Quarterly*, 77(2): 210–227.

Milić, A. (1994). *Women, Politics, Family*. Beograd: Institute for Political Studies.

Milner, A. and Browitt, J. (2002). *Contemporary Cultural Theory*. Crows Nest, Australia: Allen & Unwin.

Nemanjić, M. and Spasić, I. (Eds.). (2006). *Legacy of Bourdieu: Lessons and Inspirations*. Belgrade: IFDT/Institute for Research of Cultural Development.

Ostrower, F. (1998). The arts as cultural capital among elites: Bourdieu's theory reconsidered. *Poetics*, 26(1): 43–53.

Oxford Poverty and Human Development Initiative. (2019). *Global Multidimensional Poverty Index 2019: Illuminating Inequalities*. Available at: www.researchgate.net/publication/337474663_Global_Multidimensional_Poverty_Index_2019_Illuminating_Inequalities/citation/download (accessed on 6 March 2020).

Pakulski, J. and Waters, M. (1996). *The Death of Class*. London, UK: Sage.

Peterson, R.A. (1997). The rise and fall of high brow snobbery as a status marker. *Poetics*, 25: 75–92.

Peterson, R.A. (2005). Problems in comparative research. The example of omnivorousness. *Poetics*, 33: 257–282.

Portes, A. (1998). Social capital: Its origins and applications in modern sociology. *Annual Review of Sociology*, 24: 1–24.

Ruijter, E., Treas, J., and Cohen, P. (2005). Outsourcing the gender factory: Living arrangements and service expenditures on female and male tasks. *Social Forces*, 84: 305–322.

Sen, A., North, D., and Stiglitz, J. (2013). Beyond the GPD. *The Economists*. Available at: www.economist.com/feast-and-famine/2013/04/18/beyond-gdp (accessed on 6 March 2020).

Smith, G. (2013). *The System of Liberty: Themes in the History of Classical Liberalism*. Cambridge: Cambridge University Press. https://doi.org/10.1017/CBO9780511793325

Southerton, D. (2001). Consuming kitchens: Taste, context and identity formation. *Journal of Consumer Culture*, 1(2): 179–204.
Spasić, I. (2004). *Sociology of Everyday Life*. Belgrade: Institute for Textbooks and Teaching Utilities.
Spitze, G. (1999). Getting help with housework: Household resources and social networks. *Journal of Family Issues*, 20: 724–745.
Strotmann, H. and Volkert, J. (2016). Multidimensional poverty index and happiness. *Journal of Happiness Studies*, 19(1): 167–189. https://doi.org/10.1007/s10902-016-9807-0
Tomanović, S. et al. (2006). *Social Studies of Some Aspects of Social Transformation in Serbia*. Beograd: ISIFF.
Vanek, J. (1978). Household technology and social status: Rising living standards and status and residence differences in housework. *Technology and Culture*, 19: 361–375.
Warde, A. (1997). *Consumption, Food and Taste*. London: Sage.
Warde, A., Martens, L., and Olsen, W. (1999). Consumption and the problem of variety: Cultural omnivorousness, social distinction and dining out. *Sociology*, 33(1): 105–127.
Wilhelm, P. (2002). International validation of the corruption perceptions index: Implications for business ethics and entrepreneurship education. *Journal of Business Ethics*, 35(3): 177–189.
World Economic Forum. (2020). *Global Gender Gap Report 2020*. Available at: www3.weforum.org/docs/WEF_GGGR_2020.pdf (accessed on 4 March 2020).

10 Measuring Professional Life

Borislav Đukanović

10.1 Professional Life in Montenegro: Research

Family and professional life are the most important areas of everyday life and are key to describing the lifestyle of the stratum. Modern societies are societies of professions and, at the same time, the most important element of social stratification. In this context, it is especially important to know that in Montenegro the percentage of those who are permanently unemployed or are only occasionally employed is almost 30%. This fact is especially unfavorable if we keep in mind that the ratio of actively and passively employed is 52% to 48%. If it is known that the average duration of unemployment is four years and seven months, it is self-evident why the quality of life of the population is low.

The loss of jobs of those who were employed occurred through no fault of their own in two-thirds of the cases (redundancy, the employer stopped working, was fired without fault). A little over a quarter have lost their job due to personal reasons, but in these cases, it was not their fault (injuries; chronic illness; old age, although not yet reached retirement age). The number of those who lost their jobs through no fault of their own is almost negligible—about 1%. In particular, it should be noted that there are no statistically significant differences in the reasons for job loss according to class affiliation, consumption index and education. This indirectly shows that inclusion in the global labor market does not take into account the cultural, social and stratification specifics that are especially present in small countries such as Montenegro.

10.1.1 Research Method

Research design and methods are closely related to investigation of the quality of life. The sample consists of 805 respondents. The raw data was entered into the computer software program SPSS.

10.1.2 Key Findings

Regardless of whether they are looking for a job for the first time or have lost their job, unemployed people who have been out of work for years stop looking for a job in 51.2% of the cases. Why? The key reason is that the unemployed believe that informal connections are the fastest and most efficient way to get a job, and they do not have such connections. Another less significant reason is the lack of special knowledge and skills. Although a small percentage of the unemployed possess special knowledge and skills, it is encouraging that respondents as a whole possess them in relatively significant percentages: thus, 29.4% know one or several foreign languages, 33.9% work on a computer, 9.8% play an instrument or pursue another branch of art as a hobby, 39.0% have special technical knowledge and skills and 37, 1% are active drivers. These percentages are not negligible and show that the working-age population of Montenegro has some potential to be involved in global work processes, especially if we keep in mind that all these skills and special knowledge are usually acquired independently. Of course, it is even more important that the first job or a new job (for those who lost it) is less dependent on various informal relationships and much more on the competencies and abilities of job seekers.

Completely contrary to some widespread stereotypes about laziness toward work, the working population of Montenegro is extremely motivated to work even when they have a permanent job. More than half of all working respondents are willing to make significant sacrifices in order to provide the necessary means for the family to live. So more than 90% of all employees are ready to endure various, often humiliating, conditions in order to find an additional job or start a new one. The only frustrating factor for which they are less prepared is the possible loss of a job. The reason is quite acceptable because even in much more developed societies, whereas jobs are easily lost, new ones are also much easier to find than in Montenegro. The two strongest sources of the most severe stress for a person are the loss of a spouse by death and loss of a job. Even with the risk of losing their job, 38.2% of respondents would accept a profitable job in the 'gray economy', and 43% would replace a permanent job with a temporary one only if it was significantly better paid.

Although dissatisfaction with salary and the need to provide for the family with additional or sole financial resources are the main motivating factors, we should not lose sight of the fact that behind these answers are strong impulses for professional self-promotion, which is important when it comes to future efforts by Montenegro to get involved in global labor market processes. This can be seen in an indirect way from the following findings: for members of the lower social strata, the main motive for changing jobs is a higher salary, but for members of the upper strata, the motive to move to more interesting jobs is that they will provide them with greater professional satisfaction. The conclusion is self-evident; when employees achieve a satisfactory salary, the interest of the job comes to the fore, which, at the same time, provides opportunities for professional affirmation and ascent along the professional ladder. Further, the results of this research have unequivocally shown that members of the higher classes are more willing to take entrepreneurial risks than members of the lower classes. This is understandable if it is understood that members of upper classes have significantly higher average incomes, better education and a higher position on the social ladder; have significantly more special knowledge and skills (especially those necessary to start an entrepreneurial activity) and have significantly higher social capital (informal connections and acquaintances required). They also have a significantly higher cultural capital. Members of the lower classes are less willing to take risks if they have any permanent paid job. There is another important, but underemphasized, factor: members of the lower strata are much more exhausted by everyday, mostly manual, work, so that they have very little psychophysical capacity for additional work, even if they had such jobs.

When respondents were asked about three job opportunities out of those listed in order of importance, they stated the following:

1. Doing along with a permanent job additional jobs under the contract, or 16.5%.
2. Starting an independent business ('self-employment'), or 16.1%.
3. Doing every paid job, or 15.5%.

It was particularly interesting to examine the differences by stratification. No statistically significant differences were found in the first choice; in the second, there were statistical differences. Entrepreneurs and the self-employed are significantly more willing to start an independent business ($\chi^2 = 19.675$; df $=7$; $p = 0.006$). In the third choice, significant differences were also found; experts are significantly less ready to start a new job, and officials and technicians are significantly more ready to do so ($\chi^2 = 25.417$; df $= 42$; $p = 0.013$). These findings confirm that those who have already taken risks (entrepreneurs and the self-employed) are ready to take the risk again. Further, the main reason for not taking risks is the financial situation, especially for the lower social strata who, despite low wages, do not want to lose even such small salaries because by losing their job they have little chance of finding any paid job in the near future. However, members of the upper classes—that is, experts—are also not ready to take risks, because if you accept the risk, business failure can mean the loss of relatively good salaries, with the risk of not receiving a similarly good salary in a new job. Although this risk is lower than for members of the lower strata, it exists for the simple reason that the loss of a job is accompanied by considerable uncertainty in finding a new one for members of the upper strata in Montenegrin society as well as on the planetary level.

Manifesting problems at work best reflects strata differences. Problems at work are very pronounced in at least one-fifth of the employees (Table 10.1, items 1–5), whereas item 8 (nonpayment of contributions) occurs in 12% of the cases.

In addition to the fact that a quarter of employed respondents receive a salary irregularly, and more than a third do not receive financial compensation for overtime work, a fifth of the employees are denied basic labor rights (right to sick leave, right to vacation, work tasks are not clearly defined), and 12% due to nonpayment of contributions are not entitled to health and pension insurance. All these problems at work are significantly more pronounced among members of the lower than the higher strata.

Finally, employed respondents were asked about life plans for the next five years (Table 10.2).

More than half of the respondents do not have life plans in any area of life in the next five years, and when it comes to schooling plans, the percentage is more than three-quarters. Overall, this is rather discouraging. It is interesting to see to which social strata those who do not have life plans in the next five years mostly belong to. In all four areas of life, members of higher social strata plan significantly more than members of lower strata. These differences are at a zero level of significance. Why are members of the lower classes significantly more common among those who do not make life plans for the next five years? In most cases, they did not make

Table 10.1 Problems at work

No.	Question	Yes		No		Total:	
		N	Valid %	N	Valid %	N	Valid %
1	Irregular salary	134	25.2	397	49.3	531	66
2	Unpaid overtime	181	34.1	350	65.9	531	66
3	Denial of sick leave	98	18.5	431	81.5	529	65.7
4	Denial of annual leave	104	19.7	425	80.3	529	65.7
5	Undefined work tasks	107	20.2	422	79.8	529	65.7
6	Vilification	47	8.9	481	91.1	528	65.6
7	Sexual harassment	7	1.3	521	98.7	528	65.6
8	Nonpayment of contributions	63	12	464	88	527	65.5

Source: Đukanović (2018, 184).

Table 10.2 Life plans of respondents in the next five years

No.	Question	Yes N	Yes Valid %	No N	No Valid %	Total N	Total Valid %
1	In terms of career	333	45.2	403	54.8	736	91.4
2	In terms of schooling	194	26.5	539	73.5	733	91.1
3	In terms of housing (finding, expanding, arranging an apartment)	353	48.0	382	52.0	735	91.2
4	In terms of family planning (weddings, births, separations, etc.)	303	41.3	431	58.7	734	91.2

Source: Đukanović (2018, 189).

plans for the future because their similar plans in the past had failed. The only exception is the schooling plan, where no statistically significant differences were found between the strata. This is to be expected, as success in schooling mostly depends on the person himself, and not on external circumstances. As a reason for the failure of past plans, members of the lower strata in 19.2% of the cases stated that they could not plan because there were no conditions for that, and members of the upper strata said that they succeeded because there were conditions for successful planning. Members of the lower strata pointed out in a similar percentage (18.0%) that the plan was good but they simply did not have the good fortune to realize it, and members of the upper strata pointed out that the failed plans were not realistic, given the changed circumstances, thus showing a more realistic and a more critical attitude toward their planning.

10.2 Conclusion

Members of higher social strata have significantly more often a better education, they found it much easier to find a permanent and well-paid job because they significantly more often possessed the necessary special knowledge and skills and good social connections. This made it significantly easier for them to change an existing job for a new better-paid job. With significantly higher economic, social and cultural capital, they made significantly more successful life plans in the next five years, among other things, due to good experiences from previous planning.

With few exceptions, all analyzed aspects of professional life were better in the higher social strata than in the lower social strata, as discussed earlier (Bourdieu and Passeron, 1977; Bourdieu, 1986, 1987, 1990, 1997, 1998, 1999, 2003, 2005; Bourdieu, Wacquant, and Farage, 1994; Bourdieu and Wacquan, 1998; Bourdieu and Passeron, 2012)

In addition to being significantly more often employed and working in better-paid jobs, members of upper classes are significantly more willing to take entrepreneurial risks, primarily because they have specialized knowledge and skills and, therefore, greater self-confidence, but also because they have significantly higher initial financial resources for starting a business. Finally, they have significantly better social ties with the nomenclature. Having significantly more economic, social and cultural capital to start entrepreneurial activities reduces the overall risks of failure and significantly increases the success of the assessment of entrepreneurial activities in relation to members of lower classes.

Finally, we can conclude that indicators of economic capital most clearly separate more from the lower classes; then social capital (much better formal and informal connections to individuals and institutions that have great social power and reputation) and finally cultural capital (higher levels of education and special knowledge and skills) are especially important in obtaining and retaining good jobs in the labor market.

References

Bourdieu, P. (1986). The forms of capital. In J. Richardson (Ed.), *Handbook of Theory and Research for the Sociology of Education* (pp. 241–258). New York, NY: Greenwood.

Bourdieu, P. (1987). What makes a social class. *Berkeley Journal of Sociology*, 32: 1–18.

Bourdieu, P. (1990). *In Other Words: Essays Towards a Reflexive Sociology*. Stanford, CA: Stanford University Press.

Bourdieu, P. (1997). *Language and Symbolic Power*. Cambridge: Polity Press.

Bourdieu, P. (1998). The myth of globalization and the European welfare state. In *Bourdieu Acts of Resistance* (pp. 29–44). Cambridge: Polity Press.

Bourdieu, P. (1999). *Signals of Lights*. Belgrade: Zavod za udžbenike.

Bourdieu, P. (2003). The factory of economic habitation. *Acts of Research in Social Sciences*, 150(1): 79–90.

Bourdieu, P. (2005). *The Social Structures of the Economy*. Cambridge: Polity Press.

Bourdieu, P. and Passeron, J. (1977). *Reproduction in Education, Society and Culture*. London: Sage.

Bourdieu, P. and Passeron, J. (2012). Reproduction in education, society and culture. *Word*, 82(28): 67–121.

Bourdieu, P. and Wacquant, L. (1998). On the tricks of imperialist reason. *Acts of Research in Social Sciences Sociales*, 121(2): 109–118.

Bourdieu, P., Wacquant, L., and Farage, S. (1994). Genesis and structure of the bureaucratic field. *Sociological Theory*, 12(1): 1–18.

Đukanović, B. (2018). Everyday life and lifestyles of social stratas in Montenegro. In D. Vukčević et al. (Eds.), *Social Cross Section of Montenegrin Society* (pp. 102–337). Podgorica: CANU.

About the Authors

Mirjana Radović-Marković
Mirjana Radović-Marković is a full professor of entrepreneurship. She holds BSc, MSc and PhD degrees in economics. She has served as professor at a number of international universities, foundations and institutes. She has written more than 30 books and 250 peer-reviewed journal articles. For her contribution to science, she was elected fellow of the European Academy of Sciences and Arts, Salzburg, Austria, and then fellow of the Academia Europea, London, United Kingdom, Royal Society of Arts, London, United Kingdom and six others.

Borislav Đukonović
In 1965, Borislav Đukonović enrolled in the Faculty of Philosophy (Sociology Department) and attained a degree in sociology in 1967. He is a member of the Board for Social Sciences at the Academy of Sciences and Arts of Montenegro. Since 2019, he is an academic, Euro Mediterranean Academy, Athens, Greece.

Dušan Marković
Dušan Marković was born in Belgrade in 1960. He is a professor of applied studies in Belgrade. His teaching field is information technologies and systems, on the subjects designing database and business intelligence systems. His research interests are in digitization and virtualization such as virtual education, virtual production and virtual business interaction of heterogenic business systems.

Arsen Dragojević
Arsen Dragojević, MSc, was born in Šabac. He earned his bachelor's and master's degrees in psychology in Novi Sad. He is currently doing PhD studies in social psychology at the Faculty of Philosophy, Belgrade University. Since his graduation he has been applying his knowledge in the business sector in the field of business development, HR and market research.

Index

Note: Page numbers in *italics* indicate a figure and page numbers in **bold** indicate a table on the corresponding page.

Bourdieu, P. 105–109, 112–119, 120, 123–125, 132–133
business 13, 15–18, 34, 36, 38, 42, 50–56, **52**, 63, 67, 83, **84**, **85**, 101, 109, 113, 154, 157; culture 14, 19; development 18, 99, 102; Doing Business List *21*, 23, 24; environment 22–24, **23**, 64–65, 79; flows 3–4; models 20, 25; processes 16, 32, 51, 56; risk 80, 81, 128; strategy 15, 31; success 78, 82, 123

class 15, 107–108, 110, 111, 114–117, 119, 123, 125–127, 128, 131, 148, 152
company networking 32
competitiveness 3, 50, 77, 79, 80, *81*, 102, 103, 126
consumption 105, 107–112, 114, 117, 125–132, 152
corruption perception index 137, 138
COVID-19 7, 57, 80, 82, **84**, 85, 88, 91, 94, 100, 139, 147
creative education 25, 44
crisis 6, 7, 36, 50, 80–85, **84**, **86**, 88, 91–95, 100, 111, 113, 143, 147

democracy index 134, 135, *136*
deprivation 112, 128, 130, 131, 142, 148
distance learning 40–43, 49, 57

education strategy 46, 57
employee agility 33, 34

enterprise 15–16, 31–32, 40, 50–51, **52**, 53, 55–57, 63, 65, 79, 80–81, 100, 102
entrepreneurial skills 20, 25, 65, 66
entrepreneurship 3, 13–15, 17–19, 20, **21**, *22*, 24–25, 77
everyday life 105–109, 112, 117, 120–126, 128, 152

gender gap index *143*, 144, *145*
global gig 66
global health security index 138, 139, 147
globalization 3–7, 13–16, 19, 20, 21, *22*, 63, 90
globophilia 4
globophobia 4

habitus 108, 117, 133
human freedom index 145

impact 3, 4, 6, 7, 14, 15, 19, 20, *22*, 42, 48, 51, 63, 65, 66, *81*, 83, 85, 101
individuality 44–48, 123, 126
information technologies 48
internet 20, 25, 38, 40, 41, 49, 57, 64, 67, 72

Legatum global prosperity index 99, 100
lifestyle 107, 117, 126, 127, 128, 129, 130, 131, 132, 152

manager 65
Montenegro 3, **21**, 22, **23**, 24, *41*, *42*, 65, 67, 68, 72, *79*, **80**, 81–83, 93–95, 99, 100, 101, *102*, *103*, 105, 109, 111, 116, 118, 119, 120, 121, 125, 128, 132, 133, 135, 136, 137, **138**, 139, 140, 141, *142*, *143*, 144, **145**, 146, *147*, 148, 152, 153

new forms 3, 7, 25, 38, 57, 63

organizational resilience 77, 78, 82

pandemic 7, 50, 57, 80–82, 90–92, 95, 139, 147
platform 55, 57, 64–66, 68
poverty index 142
project teams 38

resilience 3, 77, 78, *79*, **80**, *81*, 82, 88, 90–95, 100, 121, 138

social progress index 140, 141, *142*
social strata 110, 113–115, 117, 119, 120, 123, 125, 128, 131–133, 153–156

trends 3, 14, 19, 40, 50, 54

university 40, 42, 50–52, *53*, 55, 56, 57, 68, 142, 146

virtual enterprises 31, 36, *53*
virtual team 34–36, 38, 54–56

Western Balkans 17, 41, 65, 67, **80**, 99, 100

For Product Safety Concerns and Information please contact our EU representative GPSR@taylorandfrancis.com
Taylor & Francis Verlag GmbH, Kaufingerstraße 24, 80331 München, Germany

www.ingramcontent.com/pod-product-compliance
Ingram Content Group UK Ltd.
Pitfield, Milton Keynes, MK11 3LW, UK
UKHW021425080625
459435UK00011B/157